QUALITY ASSURANCE

METHODS MANAGEMENT AND MOTIVATION

Dr. Hans J. Bajaria
Editor

Published by:
Society of Manufacturing Engineers
Marketing Services Department
One SME Drive, P.O. Box 930
Dearborn, Michigan 48128
In cooperation with:
American Society for Quality Control
161 West Wisconsin Avenue
Milwaukee, Wisconsin 53203

QUALITY ASSURANCE
METHODS, MANAGEMENT AND MOTIVATION

Copyright 1981 by the
Society of Manufacturing Engineers
Dearborn, Michigan 48128

First Edition

All rights reserved including those of translation. This book,
or parts thereof, may not be reproduced in any form without
permission of the copyright owners. The Society does not, by
publication of data in this book, ensure to anyone the use of such
data against liability of any kind, including infringement of
any patent. Publication of any data in this book does not
constitute a recommendation of any patent or proprietary right
that may be involved.

Library of Congress Catalog Card Number: 81-50392

International Standard Book Number: 0-87263-067-6

SME and ASQC wish to express their acknowledgement and appreciation to the
following publications for supplying the various articles
reprinted within the contents of this book.

Chemical Engineering Progress
American Institute of Chemical Engineers
345 East 47th Street
New York, New York 10017

Industrial Engineering
American Institute of Industrial Engineers
25 Technology Park/Atlanta
Norcross, Georgia 30092

Iron Age
Chilton Company
Chilton Way
Radnor, Pennsylvania 19089

Journal of Engineering for Industry
American Society of Mechanical Engineers
345 East 47th Street
New York, New York 10017

Journal of the British Nuclear Energy Society
The Institution of Civil Engineers
P.O. Box 101
26-34 Old Street
London EC1P 1JH
England

Journal of the Institute of Quality Assurance
Institute of Quality Assurance
54 Princes Gate
Exhibition Road
London SW7 2PG
England

Precision Metal
614 Superior Avenue West
Cleveland, Ohio 44113

Production Engineer
Journal of The Institution of Production Engineers
Rochester House
66 Little Ealing Lane
London W5 4XX
England

The Quality Reliability Journal
Indian Association for Quality and Reliability
c/o SQC Unit, ISI 4
Richmond Road
Bangalore 560 025
India

RCA Engineer
RCA
Building 204-2
Cherry Hill, New Jersey 08358

Tooling & Production
Huebner Publications, Inc.
5821 Harper Road
Solon, Ohio 44139

Union of Japanese Scientists and Engineers
5-10-11 Sendagaya
Shibuyaku
Tokyo
Japan

Grateful appreciation is also expressed to:

Institute of Electrical and Electronic Engineers, Inc.
345 East 47th Street
New York, New York 10017

Society of Automotive Engineers, Inc.
400 Commonwealth Drive
Warrendale, Pennsylvania 15096

PREFACE

Quality of products and services are more in focus now than they have ever been as a strategic element in maintaining the U.S. leadership in technology and business. Some of the external factors that are making quality considerations important are: foreign competition, consumer organizations, product liability issues and consumer attitudes. Equally important are the internal factors such as excessive costs of poor quality, quality of worklife issues and a fear of loss of reputation.

This book is a collection of papers to express modern thoughts on quality assurance and to highlight its place within everyday business activities. This book also contains different work elements that are part of good quality assurance practices.

Chapter 1 discusses modern quality assurance as it is understood today.

Quality assurance organization and management are key elements in the total organizational structure that influence product or service quality. Chapter 2 highlights quality organizational and management considerations.

Methods of quality control include evaluation of incoming materials, process reliability, capability and performance studies, process controls and production problem solving by statistical techniques. Chapter 3 outlines these methods.

Quality assurance auditing has become a necessary element in monitoring conformance to requirements and/or fitness for use at any stage. Audits are done within plants using quality assurance activities to provide independent assessment. Audits are also done by quality staff in the various plants, and by original equipment manufacturers in the vendor's plants, etc. Since prevention measures do not always exist, it has been traditionally popular to audit the products and determine whether it is an acceptable outcome or not. Auditing has developed into a full-fledged discipline. Chapter 4 covers various elements of quality assurance auditing.

Chapter 5 covers the four essential elements: prevention, assessment, internal failure and external failure that are generally categorized under the title "quality costs".

Attitudes of personnel that are part of producing products and services are extremely important in trying to achieve quality and reliability goals. Chapter 6 covers motivational programs that have been tried with differing degrees of success.

Chapter 7 identifies how quality affects productivity.

Chapter 8 covers those issues that link quality and reliability of products.

Quality assurance has been an important consideration in minimizing and/or eliminating product liability exposure. Chapter 9 emphasizes these articles that examine the relationship of quality assurance and product liability.

Quality assurance practices differ from industry to industry. In some industries, they may consist entirely of inspection-related activities, whereas, in some others, they may include prevention and inspection in varying proportions. Chapter 10 examines the state-of-the-art of quality assurance in different industries.

One of the key elements of U.S. leadership in the world is based on its technological innovations in products and services. In recent years, it has become apparent that the U.S. is not presently maintaining as big a lead as before in quality and productivity. Chapter 11 examines different issues that are at stake in international markets with respect to product quality.

The 1980's are identified as a decade in which America will try to achieve energy independence and also regain its strength in certain segments of the market in which it has lost significant market share, due to a lag in quality and productivity. Chapter 12 looks at recent issues and trends that will determine the future of the U.S. and its actions.

I wish to express my gratitude to the authors whose works appear in this volume. I also express my thanks to the publishers who generously allowed us to use their material. They are: *RCA Engineer, Iron Age, Precision Metal, Tooling & Production* and *Production Engineer*.

Special acknowledgement should go to the Society of Automotive Engineers, the Institute of Electrical and Electronic Engineers, the Indian Association for Quality and Reliability, the Institute of Quality Assurance, the Union of Japanese Scientists and Engineers, the American Society of Mechanical Engineers, the British Nuclear Energy Society, the American Institute of Industrial Engineers, and the American Institute of Chemical Engineers. My thanks is also extended to Bob King and Judy Stranahan of the SME Marketing Services Department for their efforts in producing this volume. Finally, I wish to thank my family for their tremendous patience which allowed me to make this voluntary contribution.

Dr. Hans J. Bajaria
Editor

SME

The informative volumes of the Manufacturing Update Series are part of the Society of Manufacturing Engineers' effort to keep its Members better informed on the latest trends and developments in engineering.

With 50,000 members, SME provides a common ground for engineers and managers to share ideas, information and accomplishments.

An overwhelming mass of available information requires engineers to be concerned about keeping up-to-date, in other words, continuing education. SME Members can take advantage of numerous opportunities, in addition to the books of the Manufacturing Update Series, to fulfill their continuing educational goals. These opportunities include:

- Chapter programs through the over 200 chapters which provide SME Members with a foundation for involvement and participation.
- Educational programs including seminars, clinics, programmed learning courses and videotapes.
- Conferences and expositions which enable engineers to see, compare, and consider the newest manufacturing equipment and technology.
- Publications including Manufacturing Engineering, the SME Newsletter, Technical Digest and a wide variety of books including the Tool and Manufacturing Engineers Handbook.
- SME's Manufacturing Engineering Certification Institute formally recognizes manufacturing engineers and technologists for their technical expertise and knowledge acquired through years of experience.

In addition, the Society works continuously with the American National Standards Institute, the International Standards Organization and other organizations to establish the highest possible standards in the field.

SME Members have discovered that their membership broadens their knowledge throughout their career.

In a very real sense, it makes SME the leader in disseminating and publishing technical information for the manufacturing engineer.

TABLE OF CONTENTS

1 MODERN QUALITY ASSURANCE

PLANNING AND MAINTAINING PRODUCT QUALITY
by *R.E. Fromson* and *J.N. Brecker*
Presented at the Society of Manufacturing Engineers Ohio Valley Conference, October 1977 3

QUALITY ASSURANCE PLAN—KEY TO PRODUCT INTEGRITY
by *Chas. A. Mills*
Reprinted by permission from the 1977 ASQC Technical Conference Transactions, Philadelphia 7

THE ROLE OF QUALITY TECHNOLOGY
by *Richard A. Freund*
Reprinted from ICQC '78 Tokyo Proceedings, JUSE by permission 10

2 QUALITY ASSURANCE ORGANIZATION AND MANAGEMENT

A STRATEGY FOR IMPROVING PRODUCT QUALITY THROUGH QUALITY AWARENESS AND PARTICIPATIVE MANAGEMENT
by *C.R. Davis*
Reprinted with permission © 1977, Society of Automotive Engineers, Inc. 17

A MANAGEMENT SYSTEM FOR PRODUCT ASSURANCE
by *Philip J. Birbara*
Reprinted from Chemical Engineering Progress, May 1976 27

QUALITY MANAGEMENT THROUGH QUALITY INDICATORS: A NEW APPROACH
by *Subhash C. Puri* and *John R. McWhinnie*
Reprinted by permission from the 1979 ASQC Technical Conference Transactions, Houston 34

QUALITY ASSURANCE AND OTHER DEPARTMENTS
by *David J. Murray*
Reprinted Courtesy of The Institute of Quality Assurance, London, England, from the Journal of the Institute of Quality Assurance, Volume 2, Number 4, December 1976 41

DEVELOPING A QUALITY ASSURANCE PROGRAM
by *John H. den Boer*
Reprinted from Tooling & Production, June 1980 47

3 METHODS OF QUALITY CONTROL

METHODS OF QUALITY CONTROL
by *Hans J. Bajaria*
Reprinted by permission of the author 53

4 QUALITY ASSURANCE AUDITING

QA AUDITING: A MANAGERIAL APPROACH
by *Robert E. Dellon*
Reprinted by permission from the 1978 ASQC Technical Conference Transactions, Chicago 61

THE QUALITY ASSURANCE SYSTEM AUDIT: WHO NEEDS IT?
by *Albin J. Rzeszotarski*
Reprinted from ICQC '78 Tokyo Proceedings, JUSE by permission 65

QUALITY AUDITS IN SUPPORT OF SMALL BUSINESS
by *Walter Wilborn*
Copyright 1979 American Society for Quality Control, Inc. Reprinted by permission from Quality Progress, April 1979 71

PATROL INSPECTION: GUIDE TO PRODUCTIVITY & PROFIT
by *James E. Cooper*
Reprinted by permission from the 1977 ASQC Technical Conference Transactions, Philadelphia 74

5 QUALITY COSTS

USING QUALITY COST ANALYSIS FOR MANAGEMENT IMPROVEMENT
by *Lee Blank* and *Jorge Solorzano*
Reprinted from INDUSTRIAL ENGINEERING, February 1978 79

6 QUALITY MOTIVATION

SHORTCOMINGS OF CURRENT MOTIVATIONAL TECHNIQUES
by *E.F. Thomas*
Reprinted by permission from the 1977 ASQC Technical Conference Transactions, Philadelphia 87

INTEGRATION OF TQC AND MOTIVATION PROGRAMS
by *B. Veen*
Reprinted by permission from the 1977 ASQC Technical Conference Transactions, Philadelphia 97

ZD MOVEMENT IN BHARAT ELECTRONICS LTD.
by *K.M. Shanmugam*
Copyright 1975 The Indian Association for Quality & Reliability. Reprinted by permission from The QR Journal, January 1975 104

SOME TRENDS IN PRODUCTIVITY AND QUALITY OF LIFE
by *W. Maurice Kaushagen*
© 1976 IEEE. Reprinted, with permission, from IEEE ENGINEERING MANAGEMENT REVIEW, Vol. 4, No. 4, p. 4, December 1976 114

APPLICATIONS OF PSYCHOLOGY TO QUALITY CONTROL
by *Paul Peach*
Copyright 1975 The Indian Association for Quality & Reliability. Reprinted by permission from The QR Journal, September 1975 121

7 QUALITY ASSURANCE AND MANUFACTURING PRODUCTIVITY

QUALITY AND PRODUCTIVITY—INTERACTIONS
by *H.J. Bajaria*
Reprinted by permission from the 35th ASQC Midwest Conference Transactions, Tulsa 131

QUALITY & PRODUCTIVITY—HAVE YOUR CAKE & EAT IT TOO!
by *C.R. Edmonds*
Reprinted by permission from the 1977 ASQC Technical Conference Transactions, Philadelphia 135

ON AN AXIOMATIC APPROACH TO MANUFACTURING AND MANUFACTURING SYSTEMS
by *N.P. Suh*, *A.C. Bell* and *D.C. Gossard*
Reprinted from the Journal of Engineering for Industry, May 1978, Vol. 100, © The American Society of Mechanical Engineers 141

8 QUALITY ASSURANCE AND RELIABILITY

ACTIVITIES AFFECTING QUALITY DURING ENGINEERING
by *G.F. Dilworth*
Reprinted by permission from the 1977 ASQC Technical Conference, Philadelphia 151

QUALITY ASSURANCE IN DESIGN
by *J.R.D. Jones* and *P.J. Cameron*
Reprinted by permission of the British Nuclear Energy Society from the Journal of the British Nuclear Energy Society, April 1976, Vol. 15, No. 2, pgs. 173-176 159

9 QUALITY ASSURANCE AND PRODUCT LIABILITY

QUALITY ASSURANCE AND PRODUCT LIABILITY
by *Richard B. Lynch*
Presented at the Society of Manufacturing Engineers Midwest Conference, October 1975 167

PRODUCT DEFENSE—A TEAM EFFORT
by *C.W. Walton*
Reprinted with permission © 1975, Society of Automotive Engineers, Inc. from Automotive Engineering, May 1975 179

PREPARING TOMORROW'S DEFENSE TODAY
by *Norman G. Sade*
Reprinted by permission from the ASQC PLP/76 Conference 182

10 QUALITY ASSURANCE IN DIFFERENT INDUSTRIES

QUALITY CONTROL FOR INVESTMENT CASTING WAXES
by *Myron Koenig*
Reprinted from PRECISION METAL, June 1980 189

QUALITY ASSURANCE SYSTEM FOR THE INTEGRATED CIRCUIT
by *Masahide Takanashi*
Reprinted from ICQC '78 Tokyo Proceedings, JUSE by permission 191

PRACTICAL EXPERIENCES IN ESTABLISHING SOFTWARE QUALITY ASSURANCE
by *Edward I. Keezer*
© 1973 IEEE. Reprinted, with permission, from Record 1973 IEEE Symposium on Computer Software Reliability, April 30-May 2, 1973, New York, NY 197

11 QUALITY–A KEY ISSUE IN INTERNATIONAL COMPETITION

PRODUCT QUALITY: THEORY AND PRACTICE IN MULTINATIONAL APPROACH
by *Fernando D. Negro*
Reprinted with permission © 1976, Society of Automotive Engineers, Inc. 207

QUALITY TECHNOLOGY—A BRIDGE TO INTERNATIONAL COOPERATION
by *Richard A. Freund*
Copyright 1980 American Society for Quality Control, Inc. Reprinted by permission from Quality Progress, February 1980 212

SOME NATIONAL AND INTERNATIONAL ASPECTS OF QUALITY
by *Richard A. Freund*
Copyright 1980 American Society for Quality Control, Inc. Reprinted with permission from Chemical Division News, Vol. XXIV, No. 1, February 1980 214

12 FUTURE TRENDS IN QUALITY ASSURANCE

PROGRAMMABLE CONTROLLERS, MINICOMPUTERS, AND MICROCOMPUTERS IN MANUFACTURING
by *J.L. Miller*
Reprinted from RCA Engineer, Vol. 21, No. 3, October/November 1975 221

A LOOK AT QC CIRLCES
by *Robert T. Amsden* and *Davida M. Amsden*
Reprint from Tooling & Production, June 1980 225

QUALITY CIRCLES—JAPAN'S WAY TO BETTER QUALITY
Reprinted from PRODUCTION ENGINEER, November 1979 issue "Quality Circles—Japan's Way to Better Quality" 230

NEW TECHNOLOGIES PUT QC ON THE PRODUCTION LINE
by *Robert R. Irving*
Reprinted by the Society of Manufacturing Engineers from IRON AGE, May 26, 1980; Chilton Company; 1980 231

COMPUTER-ASSISTED QUALITY CONTROL
by *John R. Coleman*
Reprinted from Tooling & Production, June 1980 235

OPERATOR'S INVOLVEMENT IN COMPANY QC PROGRAMME
by *Ravi L. Kirloskar*
Copyright 1977 The Indian Association for Quality & Reliability. Reprinted by permission from The QR Journal, September 1977 239

INDEX 245

CHAPTER 1

MODERN QUALITY ASSURANCE

Planning And Maintaining Product Quality

By R.E. Fromson and J.N. Brecker
Westinghouse Electric Corporation

INTRODUCTION

In order to assure quality components, assemblies, and processes, three engineering functions -- Design, Manufacturing, and Quality Assurance -- must work together in the planning stages of product design, tooling, and inspection/test. Each in turn must relate to each other's responsibility so as to create a total system which will have high reliability.

DESIGN CONSIDERATIONS

If one were to assume that this cooperation must start at the engineering drawing, he would be wrong. Once the drawing is on paper, changes are difficult to accomplish. Superior plans can be developed when pre-layout conferences examine alternatives, decide on preferred materials of construction, select tolerancing systems, establish gauging and testing methods, develop sub-assembly and assembly configurations, and review customer requirements. All applicable specifications should be identified.

When cost effective decisions have been reached, then drawings can be made under the direction of the design engineer. During the drawing phase, the manufacturing engineer should monitor progress for equipment, tooling and quality factors. How flat is flat? How round is round? What surface finish is needed? What tolerance has to be defined? How will it be made? Will process capability be adequate for the tolerance? Have AQL levels for vendors been established? Has in-house functional expertise been brought-to-bear on facilities, processes, gauges, test equipment and lab facilities? No drawing should be official until it is signed off by a senior manufacturing engineer.

Experience dictates that the customer's acceptance criteria be duplicated in the supplier's plant, i.e., if the buyer has a functional test, the seller is well advised to have an approved duplicate of such a test. It is not enough that the product manages to reach the shipping floor. Yield is not only a measure of profitability, but it is also a measure of field rejection experience. When defect rates are high, some components will slip through in spite of inspection. Poor field performance can mean loss of goodwill and business as well as high warranty costs.

The questions, the planning effort, and the combined effort outlined above appear to be time consuming and expensive. Those who have tried both short-cut methods and the cooperative system outlined above can vouch for the cost and time savings which can be achieved by investing in the "front end". Changes in product design, tooling and gauging after production has commenced is much more expensive, and the time factor at this point is frequently intolerable.

PROCESS/INSPECTION CONSIDERATIONS

Manufacturing of fabricated components, chemical product, and metallurgical products should all be considered processes. The specifications can take different forms, i.e., drawings, roll schedules, process sheets, etc., but basic process control procedures apply universally.

Measurements at each step in the manufacturing operation should exhibit a bell-shaped or acceptably skewed bell-shaped distribution. The measurement and display of information should be as close to the operator as possible and the operator should be held responsible for building quality into the product. It cannot be inspected into the product.

In all systems, there are some attributes that are more critical to product performance than other attributes. It pays to plan the processing of critical parameters with greater care and to monitor these parameters more closely in the shop. When concentricity is important, process from one chucking. Tolerances should be checked with process capabilities. If surface roughness is called out, are orientation, stylus tip radius, cut-off length, and traverse length specified? Are appropriate dampening measures incorporated in the tooling?

Controlling quality in the factory requires a thorough analysis of the product. In the case of processes, the process parameters can be extremely critical. One process in Westinghouse can be cited for which there is no destructive or non-destructive test which will reliably predict the product suitability for further use. The only assurance of quality that can be realized is through the simultaneous control of many process parameters within specific limits.

Fortunately, most components and assemblies can be measured. Who measures, how he measures, how frequently he measures, and when he blows the whistle are critical to reliability. In many cases it is not enough to stay within the drawing tolerance. For example, a shaft and bearing call for 0.003 inch tolerance. If the grinder operator stays to the high side (he can always take off more), and the boring machine operator stays to the low side (he, too, can always take off more), the shaft will obviously be too tight in the bearing. This occurs in spite of the fact that both components are within the drawing. If the components are produced under different line foremen, and assembled under a third foreman, all the ingredients for conflict are in place. The only solution is to have a normal distribution for both components. Control charts are excellent tools, and operator training can be most helpful.

Of course, one can mechanize and incorporate feedback in the machine control cycle to maintain normal distribution through automatic tool adjustment, but not every shop has the volume to justify such measures.

The responsibility for stopping production is usually that of the quality control representative. This is unfortunate. He should be the last one in the chain. The operator should be trained to stop producing when he recognizes trends. The foreman is next, since he should be aware of what is going on. The roving inspector is next, and the quality control department station is last in the chain. By the time he catches something, costs can be significant.

INTEGRITY ASSURANCE

Quality integrity is the result of line and staff dedication to the control of incoming material, control of manufacturing operations, control of tooling and equipment, and control of inspection gauges, instruments and testing facilities.

Every-day experiences in the real world would indicate that there is a limit to the degree of integrity one can attain. This limit can be broken by ingenuity, the combined ingenuity of design, manufacturing and quality control engineers.

The design engineer can be creative in his design. He can minimize tolerance build-up, and make the design less sensitive to variations in components. The manufacturing engineer can select processes and process controls with high reliability. The quality control engineer can choose gauges and instruments that monitor the critical aspects of dimensions and tolerances.

Very often all three functions will have to resort to non-traditional approaches. The quality control function offers a case-in-point. Since higher integrity levels can be attained with instrumentation having higher discrimination and better quantification, why not insist on better tools? If industry sits back and is satisfied with what is available, industry will have to be satisfied with higher factory costs and higher field service costs.

Of course, the established instrument maker is the obvious source for improvements. Unfortunately, this does not always prove fruitful. Your own organization could have the expertise necessary to fill your needs. If this is not the case, either the National Bureau of Standards or national technical societies can be resources to industry.

Within the Society of Manufacturing Engineers, the Non-Traditional Machining Committee was faced with industry requests for characterizing EDM and ECM surfaces in the same context as surfaces generated by traditional chip-cutting processes. Stylus-based instruments did not fit the bill. As a result of the findings of an ad hoc committee, it was possible to provide the impetus for the design of an entirely new surface texture instrument which should be much more useful than stylus instruments. The new instrument is based on the capacitance principle; it makes an instantaneous area assessment of the void fraction at the crest line. For surface texture the probe utilizes a conforming sensor and if a rigid sensor is employed, flatness can be ascertained. See Figure 1.

This case has been cited to illustrate the possibilities for expanding capabilities beyond the state-of-the-art. It is not a mission impossible.

SUMMARY

   Product quality starts at the top. Management must charge design, manufacturing, and quality control engineering with providing a plan which is integrated across functions. The functional engineers must design attainable quality into the components, tooling and gaging; the line and staff must maintain process control, starting with the operator; the functional engineers must not be limited by the state-of-the-art. Quality cannot be inspected into a product, why not build quality in?

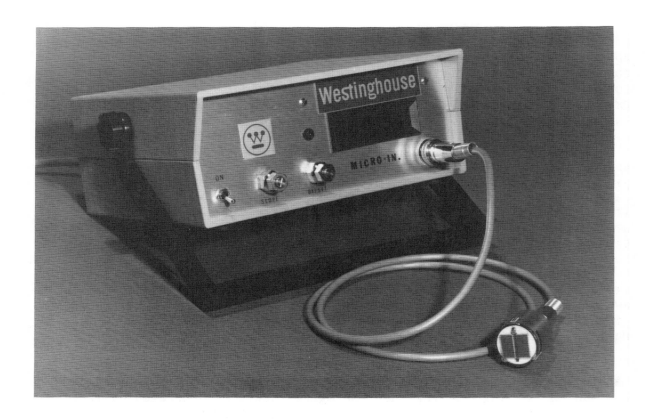

Fig. 1. Surface texture instrument.

Quality Assurance Plan—Key To Product Integrity

Chas. A. Mills
Manager, Quality Assurance
Electronic Systems Division
Westinghouse Canada Limited

Introduction

A Quality Assurance program is an integrated program of all functions within the organization which contribute to the product or service fulfilling its expected requirements. It thus makes the responsibility for Product Integrity one of the objectives assigned to management of each operating department.

The splintering of the overall objective of Product Integrity into individual objectives related to each operating department, comes very close to the old addage that what is everybody's responsibility quickly becomes nobody's responsibility.

It is essential that a Quality Assurance program is very closely integrated throughout the organization to ensure that all parties are contributing and that no responsibilities fall into the cracks between functions. A product oriented Quality Assurance Plan can serve as the means of integrating the various efforts into the single objective of assuring Product Integrity.

A Quality Assurance Manual will not ensure this team activity since it will be largely devoted to the overall policies and procedures. A product Quality Assurance Plan can zero in on the details applicable to the particular product. The Plan will, therefore, serve as the specific guidelines for all the line operations, as well as providing a base line against which performance can be measured. Therefore, the Quality Assurance Plan must cover the total life of the product or service, i.e. from its conception until the end of its life.

The Quality Assurance Plan has long been a requirement of the Canadian Military Specification, DND 1015, and more recently of the CSA Procurement Standard, Z299.1. Both of these documents refer to an organization operating a total Quality Assurance program.

The Quality Assurance Plan

In utilizing a Quality Assurance Plan over a number of years on both military and industrial products, we have found the Quality Assurance Plan to be extremely valuable. Over the same period, we have experimented with a variety of formats in an endeavour to:

1. Improve the communications.

2. Improve the clarity of the statements.

3. Simplify the preparation and use of the document.

In developing products in an industrial environment, it becomes apparent that functionally there are only a limited number of functions involved with Product Integrity regardless of the complexity of that product. These comprise:

1. Marketing.

2. Design.

3. Materials - Procurement and Control.

4. Manufacturing.

5. Application or Customer Service.

These five functions are totally responsible for assuring Product Integrity. Other functions such as Quality Assurance, Metrology, etc. are means of ensuring that these responsibilities are properly discharged.

If we examine each of these five functional areas in terms of the information they require from a Product Quality Assurance Plan, the justification of this somewhat heretical statement will become apparent.

The MARKETING portion of the Quality Assurance Plan should clearly indicate those policies and procedures which are to be used for that specific product, in order to ensure that the customer or product requirements are known to all responsible functions. Any requirements for contract review by Quality Assurance should also be clearly stated.

The DESIGN aspects of the Quality Assurance Plan should clearly define those specific policies, procedures, and controls necessary to product capability. These should include such items as documentation and drawing controls, model building, model verification tests, reliability analyses, design reviews, engineering trials, etc. In defining these specific requirements, particularly verification tests, care must be taken to ensure that types of tests are specified rather than the parameter limits. This is essential to eliminate any ambiguities and/or errors resulting from double specification.

The MATERIAL aspects of the Quality Assurance Plan should define the specific policies, procedures, and special requirements which apply to the procurement, receiving, incoming inspection, storage, and general handling of material necessary to ensure the proper materials are available for the product. Typical of the requirements which should be clarified are - availability of purchase orders to a customer, company source inspection requirements, supplier approvals, quality terms and conditions for the purchase order, objective evidence of material conformance to be obtained from the supplier, special material controls such as batching, integrated circuit screening, etc.

The MANUFACTURING aspects of the Quality Assurance Plan should cover the policies, procedures, and special requirements for the manufacture, inspection, test, pack, and shipping necessary to demonstrate that the product is and continues to be SUITABLE FOR ITS INTENDED PURPOSE. Typical of the areas which should be specified are - special process restrictions (welding approvals, etc.), special product controls, critical inspection points, sampling plans which can be used, testing program, production verification testing, pack verification tests, customer participation, etc.

The APPLICATION/CUSTOMER SERVICE aspects of the Quality Assurance Plan should clearly indicate the specific policies, procedures, and special requirements which apply to the particular product once it has been shipped to ensure that it remains SUITABLE FOR ITS INTENDED PURPOSE. This would include such items as handbooks, maintenance manuals, etc., customer service bulletins, installation instructions, de-commissioning instruction, special warranty requirements, etc.

Although the above five sections of the Quality Assurance Plan will provide visibility on all of the necessary line functions, certain of the assessment and/or support functions should also be included. Under this heading would come:

1. Quality Assurance - The Quality Assurance Plan should indicate any Quality Assurance functions which will be specific to that product.

2. Non-Conforming Material - The Quality Assurance Plan should clearly indicate how non-conforming material shall be identified, segregated, and given disposition with respect to that particular product.

Our experimentation has shown that one of the more effective ways of communicating the requirements of the Quality Assurance Plan is to have a basic Quality Assurance standard which can form the reference for all products. After developing a number of individual product Quality Assurance Plans, the format of this basic standard will rapidly materialize.

We have found that the preparation of such a standard permits the individual Quality Assurance Plan to consist primarily of the index of the main Product Assurance Plan, with each section shown as applying "in toto", or with exceptions. With this approach, the index forms the first paragraph of the Quality Assurance Plan. The first exception forms the second and so on. This abbreviated data, combined with a detailed flow diagram showing the sequence of the various control events, provides ready visibility of any non-standard requirements. It is thus fairly simple to prepare, and hence is ready on time It will not involve repeated reading of standard requirements, but rather permits the ready review of exceptions.

This type of Q.A. Plan provides an excellent reference document for determining Quality Audit requirements.

Conclusion

We have found that this type of Quality Assurance Plan brings the various line functions into an integrated program in ensuring that the products shipped are SUITABLE FOR THEIR INTENDED PURPOSE. We have also found that assigning of responsibility in this way has given individual departments an improved incentive for ensuring their operations meet the product requirements.

This approach to Quality Assurance Plans has been evolved and used as part of a Quality Assurance program certified by the Canadian Department of National Defence as meeting the requirements of their standard DND 1015. In this connection, it has been used successfully with the various NATO specifications. The economic validity of the system has been proven on a wide range of industrial products.

The Role of Quality Technology

by Richard A. Freund
Senior Staff Consultant, Quality Assurance
Kodak Park Division
Eastman Kodak Company

Introduction

There are several obvious channels for international cooperation in quality. These include close liaison between major quality control associations such as the Union of Japanese Scientists and Engineers (JUSE), the European Organization for Quality Control (EOQC) and the American Society for Quality Control (ASQC); the International Academy for Quality (IAQ); and international standards associations such as the International Organization for Standardization (ISO) and the International Electrotechnical Commission (IEC). This cooperation may take the form of exchanges of technology at major international conferences such as the one we are sharing today in Tokyo; in the form of the exchange of published or unpublished papers or books; or in the form of direct exchange between those with considerable application experience in specific areas and those having less working knowledge in these. In many situations in the standards world, committees bring together experts to share their knowledge and to develop sound common approaches.

Quality

One of the first steps in achieving standardization is to define terms, and I feel that this is appropriate here. "Quality" means different things to different people. Since this is a conference for experts in the assurance and control of quality, there should be general agreement that quality represents something that can be measured and evaluated. If it is not tangible, then it must remain elusive. It may be satisfactory to some to deal in generalities, but not to those who have the responsibility for maintaining quality standards.

The generally accepted definition of quality is: "The totality of features and characteristics of a product or service that bear on its ability to satisfy given needs". This definition implies that quality relates to needs, and includes many facets, not the least of which is the reality of sound economics. There are other interpretations of "quality", often used in a descriptive sense. These are important, but unless there is a definitive means of measure, there is no valid basis to achieve control. In effect, these can be considered more as indicators of relative quality which reflect attitudes, not action.

Having decided what quality is, now let us consider who is responsible for quality. That becomes a little more difficult. After all, isn't quality everybody's business? Isn't quality the lifeblood of any manufacturing or service organization? Doesn't management set the quality policy for the organization? Doesn't it implement these policies, assign the resources, and determine accountability? If the answers to these questions are all "yes", and I believe they are, does that imply that quality is simply a synonym for management? Isn't this a reasonable interpretation, particularly if quality can only be achieved with proper raw materials, proper employee attitudes and performance, proper manufacturing or service processes, proper economic judgment and so on? All of these are the responsibility of management.

But what is this management we are talking about? I am referring to the policy makers. This top-level entity evaluates and establishes priorities and holds the stewardship of the organization. Generally it is composed of people who have skills in selecting the technical experts in matters of finance, production, personnel, quality and so forth, and of melding these into an effective team. Of course, there is "management" at all levels within an organization, but the scope and degree of responsibility changes according to location in the organizational structure.

While quality is the responsibility of management, and management sets the policy, there are special skills required to direct and achieve the assurance and control of quality. It is true that management is responsible for quality. It is also true that without clear management support and dedication, quality cannot be assured. But at the same time, without a well-defined quality system in place, properly directed and staffed by qualified experts, it is also unlikely that quality can be assured.

In an environment of change, whether brought about by new product or service requirements; new operational concepts, techniques or basic materials; regulations; or greater consumer sophistication and demand, it certainly is not unreasonable to expect that quality control methods must also advance. Who is responsible for developing these new approaches? Management must recognize the need and provide a favorable climate. As with any other discipline, outside pressures may furnish motivation and even a general sense of direction. However, it is the experts in the profession who bear the ultimate responsibility to develop the requisite tools and technology. Some of this will be accomplished in the research atmosphere of the academic world, and some in development groups associated with the operations. However, all of it must be tempered in the field to be fully effective.

What, then, is the part played by those of us attending this conference who are not the top policy makers? We provide the "quality technology". Policies to have any value must be implemented. Implementation involves knowing how to carry out policies. Quality must be made to happen, and it takes an organized approach to make sure that all aspects are covered.

Quality Technology

What is quality technology? It is many things. It is the development and implementation of a quality system. In a broad technical sense, it includes the management of this system. It involved the planning and detailing of all actions necessary to provide adequate confidence that a product or service will satisfy given needs. It is the operational techniques and activities which sustain a quality of product or service that meets these needs.

There are many arts and sciences included in this discipline. To manage the function effectively one must be alert to the customer needs, the manufacturing or service capabilities, the skills required, the tools and techniques available, the financial implications and, clearly, how to tie everything together. What makes the quality manager's role somewhat different from the development or production manager is that sense of skepticism and determination to challenge what is planned or being carried out in order to anticipate and prevent whatever can go wrong. Prevention is almost a basic creed. Analysis is the fundamental tool of the trade. Variability and change are a way of life, and the quality professional knows they must be accounted for and dealt with on a routine basis. One major task is to distinguish between that variation which is acceptable, and that which requires corrective action.

A large quality system can utilize the talents of many specialists. A smaller one needs the same skills, but obviously must require that each person has responsibility for several areas. From time-to-time one specialty or another becomes popular, almost viewed as a panacea, but eventually falls back into the general pool of skills. The problem with this phenomenon occurs when the initial infatuation is over. Then, too often, the real benefits are ignored along with the unrealizable oversold expectations.

Since analysis plays such a key role in quality assurance, it is not surprising to find such specialties within this discipline as:

(1) the science and application of measurement;

(2) techniques for product or systems quality audits;

(3) quality cost analysis in order to place emphasis where it should be;

(4) statistics to deal with risks and uncertainties, and to effectively utilize the information collected, which is the basic commodity of the quality operation;

(5) design analysis to evaluate the product or service in terms of satisfying the given needs, including those of the customer and of society as a whole, and in terms of whether it will be possible to actually achieve the desired goals in normal production;

(6) inspection techniques, including when and how to use acceptance sampling for incoming, in-process or finished products or materials;

(7) reliability techniques to account for the effects of time or of complex assemblies;

(8) maintainability, availability and many other related required abilities;

(9) human factors since people continue to play vital roles in quality. Making it easier to do the current job involves physical and motivational considerations.

The list could continue, and omissions are not intentional. They can all be summarized by emphasizing that the customer needs must be translated into meaningful specifications; the design carefully reviewed; facilities evaluated in terms of their capability of satisfying specifications; supplier quality monitored; product or service measured against the specifications; corrective action initiated and accomplished; and feedback recycled to create new or improved products as needed.

This is a rapidly changing field. On-line process controls; automatic gaging; increased demands for precision; new technologies, both in manufacture and in measurement; new analytical tools; multivariate statistical techniques; and so forth, all challenge the analytic capabilities of the quality engineer. But the challenges are being met as is reflected by the papers presented during this conference.

Organizational Implications

Too often organizational questions seem to conceal the true functional needs of the quality operation. It shouldn't much matter where the quality function appears on the organizational chart, provided that the principles of independent assessment and the appropriate and timely feedback of information are achieved. Obviously, all the functions must be carried out in a coordinated manner, or there will be no system. However, flexibility in structure is needed to satisfy the many types of industry and operational arrangements. The important factor is that a system exists, is understood, is complete and is effective. Styles of management differ too greatly to define the one best structure. Good communications are essential and performance is the best measure.

Standards

Now let's return to the role of quality technology in international cooperation and in particular with respect to international standards. Standards are essentially a means of communication and form a viable basis for international agreements. Trade and cooperation depend on good communications and understanding. Quality is playing an ever-increasing role in this arena. Thus, there is a need for greater involvement of quality technology.

ISO and IEC technical committees develop product standards involving testing and evaluation procedures. Many of these groups could use professional advice on statitical quality control procedures, for example, as well as on specific evaluation criteria and methodology. Some product committees seem to use sampling plans that have repealed the laws of probability. It doesn't make sense to standardize on unreality. This can only cause future trouble. Other committees recommend procedures that do not reflect many of the modern techniques which are almost second nature to us. We need to offer our assistance through our national standards bodies. As a profession we need to become more involved. It is a shame to find that statistics and other analytical tools are not considered as carefully as they should be.

ISO technical committee 69 has the capability of recommending standard sampling plans, control charts, and statistical tests. IEC technical committee 56 plays a similar role in the reliability field. Currently there are quality experts serving on these two committees, but the committees only play an advisory role to the product standards writing committees, and only when requested to do so. There are already signs of increasing work loads and the need for additional qualified help. However, this may be an appropriate time to suggest that the product committees reflect on which of their problems could benefit from the work and assistance of TC69 and TC56.

CERTICO, the certification branch of ISO, is developing international inspection systems so that product may flow more smoothly across international borders. Some quality control considerations are already being incorporated in this effort, but there is much not yet accounted for.

A new ISO technical committee has been recommended to deal with the broad area of quality assurance systems. The anticipated scope of the committee will include definitions, sampling and control techniques, and the broad recommendation of guidelines for quality assurance systems. It should be obvious that this is the province of the quality technologist. It should also be obvious that this is an important recognition of the role that quality technology can play in standards work. The question is whether or not we are prepared to meet the challenge of making this an effective effort, and of steering the work of this committee in the direction of useful service to quality throughout the world.

Conclusion

Quality is everybody's busines. It is management's responsibility, but it is achievable only with a solid base of effective quality technology. As science advances, as products and services become more complex, as world trade increases, as consumer expectations rise, as safety and environmental considerations become more important, and as natural resources become scarcer there will be greater and greater need for a professional approach to quality. We have already come a long way as is evident from the technical content of this conference, and from the products we see around us. But we must continue to develop new technieques, and to use more effectively the ones we already have. We need to share our knowledge and experience with our colleagues around the world and, happily, we see signs of this being recognized through progress in the international standards community as well as the national and international quality control societies.

Management Services Division, Kodak Park, Building 56, Rochester, New York 14650, U.S.A.

CHAPTER 2

QUALITY ASSURANCE ORGANIZATION AND MANAGEMENT

A Strategy for Improving Product Quality Through Quality Awareness and Participative Management

C. R. Davis
Buick Motor Div., General Motors Corp.

WITHOUT QUESTION the priority objective of any business producing a product today must be product quality. Whether that product is an automobile or a golf ball, the customer expects to get precisely what the product is advertised to be. He expects a perfect product with no defects. Anything short of this is categorized as poor quality. He does not expect any less from a complex product such as an automobile with some 15,000 parts than he does from a simple product like a golf ball.

The customer's demand for perfection today is perhaps greater than it has ever been before. This is due to the attention given to product quality by government and consumer groups, as well as the higher price he has to pay for the product as the result of inflation. The customer is realistic to expect a higher quality level when he is required to pay a higher price for the product. Our problem in industry is that although we have steadily improved product quality, we have not kept up with customer expectations. These expectations will continue to intensify. The challenge in industry is to successfully satisfy these expectations. This is an urgent challenge and the organizations that meet it first and most effectively will be rewarded by increased sales and profit.

In developing a strategy for improving product quality, it will be shown that quality awareness and the control of quality must begin when a product is conceived and does not end until the product is purchased by the customer. An effective Quality Awareness Program is a necessary ingredient for establishing the environment for optimum control of quality. Quality awareness can be defined as a continuous communications process which informs everyone associated with the product from concept to sale of customer expectations in product quality and where the current product falls short. It stimulates total involvement of everyone to upgrade the quality of the product to meet these expectations. The control of quality is defined as everyone's responsibility. The traditional concept of Quality Control will be referred to as Inspection in this paper. Customer expectations for product quality can only be achieved through total involvement and commitment of everyone to achieving the quality goal.

FACTORS WHICH DETERMINE PRODUCT QUALITY

Three factors determine product quality: design, manufacturing process and people. Quality awareness and the control of quality must be a way of life during the design, processing and production phases of the product cycle to consistently produce at the quality level necessary to meet customer expectations. Let us look at each of these factors.

DESIGN - The first step in the life of the product is design. This step is crucial in achieving the quality level we are looking for. Therefore, effective control of quality

---- ABSTRACT ----

The customer's demand for product perfection emphasizes the need for a more effective strategy for improving product quality. Quality awareness and the control of quality must begin when the product is conceived and should not end until the product is purchased by the customer. The control of quality must be extended to and become a way of life in the design phase and the manufacturing process phase in addition to the currently existing production phase. This can be accomplished through a participative management system involving everyone throughout the product cycle in the development of the quality goal and the quality control plan.

must start with the design of the product. Most potential product quality problems can be identified and avoided in the design phase by establishing a close working relationship between the Design Engineer and the Quality Engineer. Through this relationship they can assure that the design is compatible with the plant facilities; the design can be built at the desired quality level; and any potential mis-assemblies are designed out. The Design Engineer should be aware of Murphy's Law which states that anything which can go wrong, will go wrong. This will adversely affect quality. The Quality Engineer should point out these potential pitfalls so the design can incorporate safeguards against Murphy's Law. Designing for quality is an important step in the strategy for improving product quality.

Until I moved into the Quality Control area two years ago, my entire career had been spent as a Design Engineer. Thus I feel that I can speak with some authority on this subject. I am convinced that all Design Engineers should spend some time working in a Manufacturing Plant early in their careers. Such experience would be invaluable in helping them to design future products that could be produced at a consistently high level of quality in the plants.

MANUFACTURING PROCESS - Following the design of a product, Manufacturing must develop the plant process for producing the product. This is another step in the evolution of a product where the control of quality must be given adequate consideration. The product that comes off the end of the line represents whatever quality level is designed and processed into it. The only way quality can be improved beyond this is for Inspection to reject those parts below the acceptable quality level. Rejected parts must then be repaired or scrapped, but either of these solutions represents a high price to pay for quality.

A manufacturing process which is marginal for quality capability may save a few dollars at the outset but this is a poor investment since repair and scrap costs will more than offset this saving. Fortunately, the lowest overall cost is provided by a manufacturing process that meets the product quality requirement. A key question is whether to pay now for a process that will consistently provide the desired quality level, or pay more later to correct the quality problems and risk the wrath of the customer.

One way to assure that the process will consistently produce at the desired quality level is through the use of the "80 Percent Concept." Most processes are designed to use 100 percent of the dimensional specification on the part. Over time the parts produced by the process begin to fall outside this 100 percent specification because of wear on the machine and tools, and possibly inadequate maintenance. The system is set up for Inspection to reject parts exceeding the specifi-

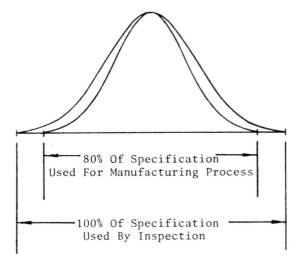

Fig. 1 - The 80 percent concept

cation, but by the time this occurs, the process is already out of control and incapable of producing parts consistently within specification. As a result, the machine must be shut down for repair to bring the process back into specification, and a large number of parts may have been produced which must be repaired or scrapped. Inspection systems are not infallible, and undoubtedly some parts get through the system and will cause quality problems with the final product and the resulting customer dissatisfaction.

The "80 Percent Concept" will prevent this situation from occurring. Simply stated under this concept the manufacturing process is allowed to use only 80 percent of the allowable specification (Figure 1). The distribution curve can shift or change without exceeding the allowable specification. The process gages are used to provide the alert when the distribution shifts or begins to exceed the 80 percent limit, and corrective action is taken immediately. Through this approach the process is never allowed to reach 100 percent of the specification. Since the Inspection gages are set to 100 percent, the system really does control quality and eliminate rejects that would require repair or scrapping.

In order to consistently produce the product quality level desired, the design and the process must be compatible with this goal. This can be accomplished through adequate quality awareness and a close working relationship during the design and processing phases among the Design Engineer, Process Engineer and Quality Engineer. It is the Quality Engineer's responsibility to see to it that the design as processed will meet the product quality goal. Many product quality problems can be eliminated when proper emphasis is placed on the control of quality in the processing phase. Processing for quality is another important step in our strategy for improving product quality.

PEOPLE - People are the third factor in determining product quality and are the most significant element in the entire life cycle of a product. Quality awareness communication encourages people involvement in the control of quality. People are the key element in the Design, Processing and Production Phases of the product cycle and determine how effectively the design and process accomplish the quality goal and provide the consistent level of product quality the customer expects.

People can negatively or positively influence the quality of the product. Unfortunately, when we think of the Production Phase of the product cycle, we tend to think of production people as being a negative influence on the quality of the product. When we experience a product quality problem, we tend to blame the production operator. Our natural reaction is that if each individual did his job properly, quality problems would not exist. This is in fact a true statement, but in most cases the criticism is leveled at the production operator who may not be the real culprit. There are strong indications that the average production operator really is interested in doing a quality job, but too often cannot for a whole series of reasons. For instance, the design or processing may not allow even the best operator to produce at the desired quality level. He may not be aware that he is creating a quality problem. In a highly integrated assembly line, one operator's quality problems can adversely affect another operator thus creating more serious quality problems. This could be the result of inadequate training on how to perform his operation and achieve the expected quality level. Or, it could result from lack of feedback on his individual quality performance. He may feel that management does not care about quality so why should he. There are elaborate systems in manufacturing plants for communicating data on quality performance to plant management, but very little gets to the individual operator who can really do something about it.

In our strategy for improving product quality, people are the most important factor. If the customer-expected product quality level is to be achieved, every person involved in the product cycle, from the time the product is conceived until it is purchased by the customer, must be highly motivated to provide the highest quality product possible, a perfect product with no defects. Our strategy must be geared to achieving this objective.

KEYS TO CONSISTENT PRODUCT QUALITY

In order to provide a product with a consistent high level of quality, it is necessary to establish quality goals, develop a quality control plan, and assure that everyone involved throughout the product cycle is sufficiently motivated to achieve the quality goal. The quality goal should be achieveable but challenging and require some stretching on the part of all concerned with the product.

QUALITY GOALS - To be most effective in improving quality, we should establish a quality goal at the time a new product is being defined and before the design phase begins. Top management must encourage quality goal setting and assure success of the goal by providing the environment that allows it to happen. All areas of the business must participate in establishing the product quality goal: Engineering, Quality Control, Processing, Manufacturing, Production Control, Data Processing, Purchasing, Financial, Personnel, Service and Sales. Without the participation of all departments, the quality goal will not be achieved, since conflicting individual department goals, such as schedules or costs, will prevent it. Can you imagine Financial being the sole department responsible for cost reduction? Of course not, and Quality Control cannot be the sole department responsible for quality improvement either. With all departments involved in setting the quality goal and making a commitment to achieve their portion, success is possible. If at any time during the product cycle, any department is not able to live up to its commitment, the other departments should be notified so as to develop an alternative which will assure achievement of the quality goal.

The overall quality goal is made up of each department's individual quality goals.

- Engineering's quality goal must be to design the product so as to eliminate potential quality problems in the processing and production phases.
- Quality Control's goal must be to assist Engineering in designing for quality, work with Processing on the production process to maximize quality, work with Manufacturing to correct unforeseen quality problems that develop in production, and establish and operate an inspection system that will catch all remaining product defects and prevent them from being shipped to the customer.
- Processing's goal must be to provide a manufacturing process that will consistently provide the desired product quality level, taking into account the capability of the equipment as well as the operators.
- Manufacturing's goal must be to train operators to produce at the desired quality level; to communicate the quality goal to everyone in the plant and the progress toward the goal; to provide adequate maintenance of equipment, housekeeping and safety; and to establish a priority on quality which assures achieving the quality goal.
- Production Control must assure that no shortages of material exist, and

set production schedules that can fluctuate if necessary in order to consistently meet the product quality goal.
- Data Processing must work with all departments to provide adequate data processing systems to perform their responsibilities most effectively, and assure that these systems continuously function properly.
- Purchasing must assure that all purchased parts used in the product consistently meet the product quality goal.
- Financial must assure that all expenditures to meet the quality goal are prudent, and must not unnecessarily restrict or delay expenditures required to meet the agreed upon quality goal.
- Personnel must assure that the quality goals are consistent with the existing labor agreement, hire the proper people for the job at all levels, and help establish a working environment within the plant that will assure achievement of the quality goal.
- Service must assure that any shipping damage which occurs to the product is corrected before it is delivered to the customer, and provide rapid feedback to the plant for corrective action on new problems they see in the field.
- Sales must assure that the agreed upon product quality goal is consistent with the customer's expectations on quality.

A total commitment to the quality goal on the part of everyone from top management down to the operators on the line is an absolute necessity for success.

In establishing the overall quality goals, it should not be assumed that the quality level of the competition or the quality expectations of the customer will remain at the current level. Quality expectations tend to be controlled by the highest quality level currently existing in the marketplace. The free enterprise system encourages competition in quality and every business is constantly striving for quality improvement in order to capture a larger segment of the market. The quality goal, therefore, becomes a moving target. Because of the delay between the time the product is conceived and the time it is produced, a quality goal must be set which will equal or better the competition when the product reaches the marketplace. A quality goal is essential in our strategy for improving product quality.

QUALITY CONTROL PLAN - Once each department has established its quality goals which together became one overall quality goal for the final product, a quality control plan must be developed. Each department must develop its individual quality control plan

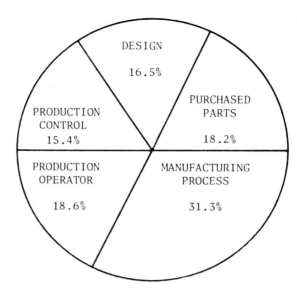

Fig. 2 - Departmental involvement in quality problems

which will assure achievement of its quality goal.

Although each department's quality control plan is different, there are similar steps which should be taken in developing a plan. The first step is to determine how potential quality problems could occur. This can be done through analysis of past experience on a similar product as well as through analysis of the new product. Typically there are many causes for quality problems with most departments in the product cycle being involved. A recent analysis of quality problems in an automotive assembly plant showed most departments shared in the responsibility for the quality problems involved as shown in Figure 2 below.

The next step is to develop safeguards to eliminate the occurrence of each potential quality problem. This can be done through design or process modifications. Every effort should be made to take the human element out of the quality problem prevention system. Dependence on the human operator as the only safeguard against a quality problem occurring is risky. The best approach is to build safeguards into the design and process to assure meeting the quality goal.

When each department has identified the potential quality problems and developed appropriate safeguards, the quality control plan (Figure 3) should be implemented on a trial basis. This is called a Production Trial Run and is a procedure whereby a small number of parts are produced under production conditions to determine if the quality control plan is adequate to achieve the quality goal. The Production Trial Run evaluates most aspects of the quality control plan: design, process, operator training and capability, buildability, purchased part quality, safe-

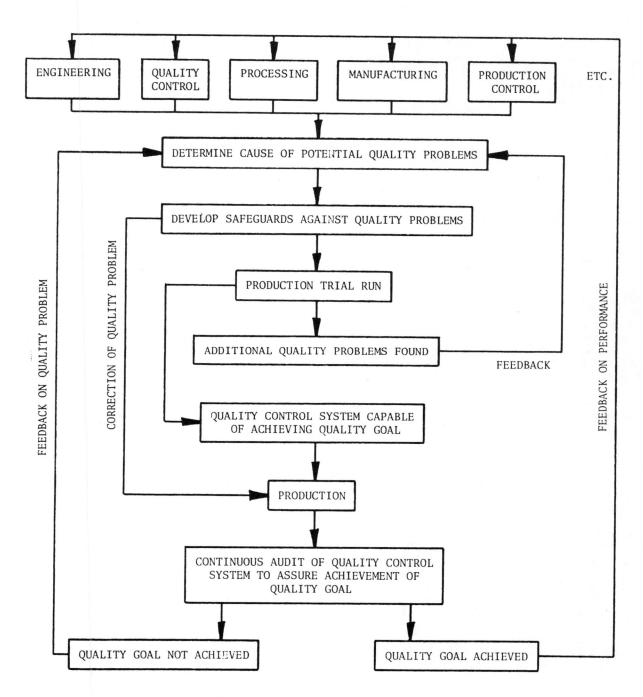

Fig. 3 - Quality control plan

guards against quality problems, and Inspection. It provides each department with an opportunity to evaluate their quality improvements. It is normal to find additional quality problems during the Production Trial Run and after all, that is its purpose. Each new quality problem must be corrected and adequate safeguards established. This procedure continues until the quality control plan is determined capable of achieving the quality goal.

Even though the quality control plan was adequate during the Production Trial Run, continuous audits of the system must be set up in production to provide the alert in the event of an unforseen problem or a failure within the system so corrections can be made. The quality control plan is a key element in our strategy for improving product quality.

MOTIVATED PEOPLE - Unless everyone associated with the product throughout the product cycle is motivated to achieve the mutually established quality goal, achievement of the goal will not be possible. Motivated people are, therefore, a key to the success of any strategy for improving product quality. The obvious questions then are: What can Management do to motivate people to achieve the quality goal? How are people motivated?

The simple answer to the first question is that Management cannot motivate its people. One person cannot motivate another person except in a very limited short term sense such as the KITA approach (or kick in the ___) described by Frederick Herzberg. The kind of motivation needed in our strategy must be long term and consistent. Motivation of this kind can only come from within each individual. In other words, self-motivation is the key.

Management per se cannot motivate people, but what it can do is to set the stage for, and indeed encourage the creation of a climate in which motivation can take root. For instance, Abraham Maslow's "Hierarchy of Needs that Motivate Behavior" tells us that individuals not only have basic physiological and safety needs, but also higher level needs for esteem and self-actualization (Figure 4). Recognizing these higher level needs, management must assure that communication lines are open and working so that each worker involved in the product cycle who can effect product quality is provided an opportunity to contribute his ideas and expertise in improving the quality of the product.

This is a very difficult thing for some managers to do since traditionally they have preferred the "top down" authoritarian style of management or "keep them in the dark" approach. This is also called Theory X by Douglas McGregor. In the world of the 1970's the authoritarian approach is becoming outdated, primarily because younger, better educated employees are not willing to work in an environment where they are told what to do and how to do it. In addition, this

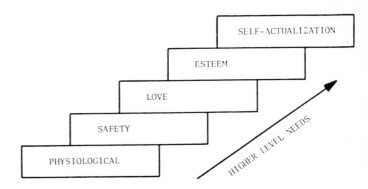

Fig. 4 - Maslow's hierarchy of needs that motivate behavior

management style often generates a negative employee reaction. Many of today's employees want to be involved.

The successful manager today uses a human relations management style, or Theory Y as McGregor describes it, which integrates individual and organizational goals. This approach encourages involvement from the bottom up in the setting of goals and the development of strategies to achieve the goal. An individual is more likely to be committed to achieving a goal he helped to establish rather than one established by someone else and given to him to accomplish. Ownership of a goal is a powerful guarantee of success. So long as the goal is consistent with management's overall objectives and the strategy does not violate company policies, management should feel very comfortable with the human relations, bottom-up, participative approach that is having increasing success today as a business management model.

Unfortunately, there are still many managers who find this approach very difficult to accept. They believe that they lose personal prestige and authority by allowing their subordinates to set the goals and strategies. In a similar manner, many Design Engineers feel uncomfortable with Quality Control involvement in the design of the product, and Process Engineers do not always readily accept Quality Control involvement in the processing. Even though they understand the gains this approach provides, the personal loss of authority is often hard to accept.

In order to understand why the participative (Theory Y) approach is the best method to provide motivation for people to improve quality, we should compare it with the authoritarian (Theory X) approach. In the past the authoritative approach was successful when the workers were on the lower levels of the Maslow Heirarchy (physiological and safety needs). They were primarily interested in survival and were willing to accept this approach to keep their job. In today's world the workers have moved much higher on Maslow's Heirarchy. They no longer fear loss of a job since the basic survival needs of food,

AUTHORITATIVE	PARTICIPATIVE
. Minimizes communication	. Maximizes communication
. Allows minimum subordinate involvement in goal setting	. Allows maximum subordinate involvement in goal setting
. Makes all decisions	. Makes use of all talent available in making decisions
. Motivates people through fear	. Motivates people through involvement
. Takes credit for results	. Recognizes subordinates accomplishments
. Defines what to do and how to do it	. Determines what should be done and how to do it jointly with subordinates
. Ignores needs of subordinates	. Recognizes subordinates' needs as a motivating factor

Fig. 5 - Characteristics of authoritative vs. participative manager

shelter and clothing are guaranteed by our welfare system. Workers are, therefore, more attuned to personal involvement, and participation which will satisfy the higher level needs of esteen, self-actualization and recognition. It is therefore becoming more difficult to motivate people to improve product quality and a participative management approach is providing to be the most effective means to that end.

The characteristics of the authoritative manager compared with those of the participative manager are shown in Figure 5. It is apparent from a review of these characteristics coupled with the knowledge of where the average worker is on the Maslow Heirarchy that motivation can only be achieved through the participative approach.

Management must be willing to accept this new approach if our quality goals are to be achieved. They must modify the traditional authoritative approach and adopt the participative approach which integrates the needs of the business (achievement of the product quality goal) with the needs of the people (involvement, participation, self-actualization and recognition).

PARTICIPATIVE MANAGEMENT CREATES QUALITY AWARENESS

The participative management philosophy is based on the premise that people take more interest in their work if they are allowed to influence decisions. People become committed to what they help to create. The higher the degree of involvement in the creation, the greater the commitment to the final product. In actuality this is saying, what is good for the employee is good for the business.

At this point you might say that the participative approach to improving quality sounds good, but does it work? The answer is yes, it does work and is working today in many large and small businesses throughout the world. It takes many forms and goes by many names such as Quality Control Circles and Participative Quality Problem Sovling. Regardless of what form the participative management approach takes, it creates a quality awareness throughout the organization which provides the environment necessary for improving product quality.

Participative management maximizes quality awareness communication related to the needs of the business that can be meshed with the needs of the people. With improvement in product quality being a top priority need of the business, the participative management approach can be used to create an awareness of the current quality position by the people and ask for their help in improving that quality position. Management is not demanding that they work harder to improve quality as would have been done under the authoritative approach, but rather is asking for their ideas, their involvement and their participation in setting quality improvement goals and in determining the most effective way to achieve those goals. Quality awareness not only involves communication of the existing quality position but also the participation of the people in determining how that quality position can be improved.

Management must work continuously to provide the environment for maintaining a quality awareness attitude within the organization. Quality awareness can become self-perpetuating within a participative management environment which encourages the workers to become involved in the communication process that provides daily information on the quality position to their peers as well as management.

This can be done through the use of closed circuit television within the plant that broadcasts information continually on the daily quality position as well as other information of interest to the people in the plant. Plant newspapers can be used to communicate quality information as well as provide recognition for those employees who have made significant contributions to improved quality. Training programs related to new models can be effectively used to perpetuate the quality awareness attitude among the people in the plant and to help establish this attitude in new people starting to work in the plant. All of these quality awareness activities should be handled by the workers and will require very little from management other than encouragement once a quality awareness attitude has been established. The workers in the plant will come up with an unending flow of quality awareness ideas for communicating on TV, in newspapers and in training programs that will effectively encourage their peers to join the quality awareness and quality improvement team.

Some skeptics may feel uneasy about turning these communications systems over to the workers, but personal experience has shown me that it works and even more effectively than I anticipated. The worker has as much stake in the success of our business as management, and the involvement in the quality awareness communications process provides the kind of self-actualization and recognition opportunities they are looking for. What they say about the company and its quality achievement has more credibility among their peers than if management said the same thing.

PARTICIPATIVE MANAGEMENT THROUGH QUALITY CONTROL CIRCLES

Quality Control Circles are an excellent example of how the participative management approach can create quality awareness, improve product quality and reduce cost while providing self-actualization and recognition for the people involved. Quality Control Circles have been in operation in Japan since 1962. They are based on the philosophy that quality is the responsibility of everyone in the organization, management and workers. The Quality Control Circle philosophy is also based on the belief that human resources are the most important resources, especially in improving product quality.

Quality Control Circles are made up of employees from the same work group who meet to discuss quality problems, investigate causes, recommend solutions and take corrective action. Membership in the Quality Control Circle is purely voluntary. All circle members receive training in quality control methods as well as problem solving techniques.

Every circle member is involved in resolving quality problems and creativity is encouraged. Since the problems are in the circle's work area, they tend to be experts on that area and their involvement in solving problems assures a commitment to make the solution successful. In addition to Japan, the Quality Control Circle approach has been used successfully in other countries including the United States.

Let us look briefly at how we might set up a Quality Control Circle or a Participative Quality Problem Solving System. We should start on a small scale and expand by building on its success. First we should locate a work group in a manufacturing plant with one or more chronic quality problems that the authoritative management system has not been able to solve. We then explain the participative approach to the first line supervisor, the workers and the Union. Management and the Union must be committed to the participative approach if we are to proceed. We then ask for volunteers to become involved. These volunteers go through a problem-solving technique training program and are made aware of all aspects of the product design and processing that relate to their work area. In short, they are provided background information as well as the tools to solve problems. Thereafter, they are continually updated on all new information relating to the product and their work area. An elaborate quality awareness communications system is set up to accomplish this. Closed circuit TV monitors throughout the plant are used to continuously broadcast data on product quality broken down by work area.

Armed with adequate data and the knowledge of problem solving techniques, the group then attacks the quality problem, determines the cause, develops a solution and implements the solution. They are well equipped to do this since they probably know more about the problem than anyone else and have developed ideas on how to correct it. When they need technical help, it is provided. As a result of this approach to quality problem solving, everyone wins. The customer gets a higher quality product. The employee has become involved in creating the solution to a quality problem which results in self-actualization. Management has been able to build a more effective working relationship with the employee while simultaneously improving product quality and customer satisfaction.

SUMMARY

In summing up our strategy for improving product quality (Figure 6), I have attempted to show that quality awareness and the control of quality must begin when the product is conceived and does not end until the product is purchased by the customer. The control of quality is the responsibility of everyone associated with the product throughout the concept-to-sale cycle.

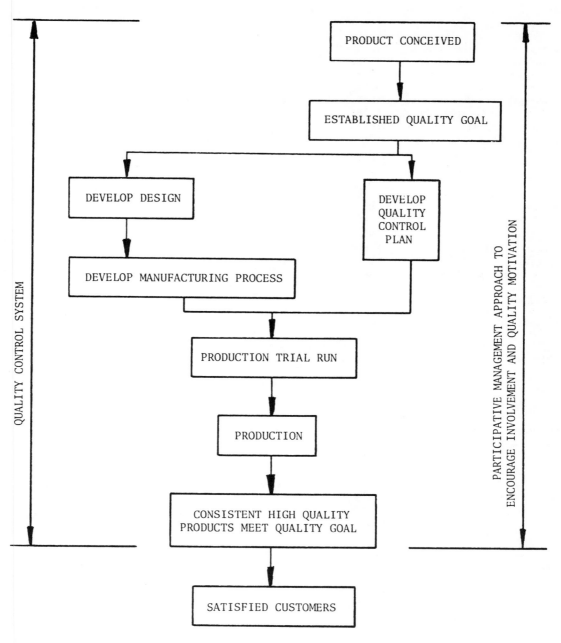

Fig. 6 - Strategy for improving product quality

The three major factors determining quality are design, manufacturing process and people. People are involved throughout the product cycle and can improve product quality most consistently through incorporation of adequate quality controls in the design and manufacturing process phases. People involved in the production phase of the product cycle can help maintain and improve product quality if they are properly motivated.

In order to assure a consistent high level of product quality, everyone involved in the product cycle must be involved in establishing a quality goal. This involvement provides the ownership of the goal that assures the kind of commitment needed for success.

A Quality Control Plan must then be developed by each department leading to achievement of its quality goal. The plan is evaluated through the use of a Production Trial Run to assure its adequacy in the real world of the plant before full production starts.

Finally the key to our strategy for improving product quality is people motivated to achieve the quality goal. This can be done most effectively through a participative management approach. An effective form of participative management is the Quality Control Circle system. Through quality awareness communication everyone in the business is encouraged to become involved and participate in our strategy for improving product quality.

Implementation of this strategy for improving product quality can be successful on any product. Only through an approach of this kind, which is based on quality awareness and involving participation of everyone in the product cycle in improving and controlling quality, will we be able to meet the urgent challenge of the customer's demand for product quality perfection.

REFERENCES

1. A. P. Chaparian, "Teammates: Design and Quality Engineers." Quality Progress, April, 1977.

2. D. R. Hampton, C. E. Summer and R. A. Webber, "Organizational Behavior and the Practice of Management." Scott, Foresman and Company, 1968.

3. W. F. Schleicher, "Quality Control Circles Save Lockheed Nearly $3 Million in Two Years." Quality, May, 1977.

4. D. McGregor, "The Human Side of Enterprise." McGraw-Hill, 1960.

5. W. A. Hill and D. Egan, "Organizational Theory, A Behavioral Approach." Allyn and Bacon, 1968.

6. C. A. Mills, "Quality Assurance Plan - Key to Product Integrity." American Society for Quality Control, 1977.

7. H. Quong, "QC Circle - Evolution or Revolution." American Society for Quality Control, 1977.

8. Y. Kondo, "Creativity in Daily Work." American Society for Quality Control, 1977.

9. K. Ishikawa, "Quality Analysis." American Society for Quality Control, 1977.

10. D. A. Amsden and R. T. Amsden, "QC Circles: Applications, Tools, and Theory." American Society for Quality Control, 1976.

11. K. Ishikawa, "QC Circle Activities." Union of Japanese Scientists and Engineers, 1968.

Reprinted from Chemical Engineering Progress, May 1976

A Management System For Product Assurance

Chemico's Reliability and Quality Assurance Dept. does not just monitor design/development activities, but joins as an active team member in finding solutions to problems in these areas.

Philip J. Birbara, Chemical Construction Corp., New York, N.Y.

A management system has been adopted by Chemico Process Plants Co. to assure delivery of a product that would meet contractual requirements within the constraints of cost, schedule, and reliability. The system has been implemented by the establishment of a Reliability and Quality Assurance (R&QA) Dept., which institutes controls to insure the attainment of product requirements by monitoring the results of these controls, correcting deficiencies, and retaining data to prevent recurring errors on subsequent projects.

In 1974, two significant developments resulted in management's decision to establish such a group. First, Chemico was acquired by Aerojet-General Corp., an organization that has pioneered in the successful implementation of R&QA techniques. And second, two sizable contracts were awarded, an ammonia complex presently being constructed within the U.S.S.R., and a demonstration plant for the liquefaction and gasification of high sulfur coal. The latter is a joint venture with Union Carbide Corp. known as Coalcon.

By establishing an R&QA organization, it was anticipated that benefits from previous experience would be fully exploited and no potential reliability problems overlooked. Minimizing errors results in costs improvements in the areas of scheduling delays, reworks, liabilities, and overall engineering manpower cost escalation.

Team approach

The use of the team approach has been a major factor in the success of past R&QA engineering programs because this technique recognizes that R&QA does not merely monitor and record design/development activities, but joins as an active team member in finding solutions to problems in these areas. The fundamental role of the organization is two-fold:

1. To ensure that all actions necessary to achieve the required system safety, reliability and quality are being taken by the respective company organizations, not as separate entities independent of each other, but rather on an integrated program-wide basis.

2. To independently evaluate using such techniques as design reviews, failure analysis, quality audits, and nondestructive testing.

For each project, emphasis is placed on the development of a planned program that provides 1) integration of reliability, quality and system safety into all aspects of design, construction, test, and manufacturing, and 2) product quality within a continuous feedback system that emphasizes design assessment, resolution of risk areas, and integration of verification controls throughout the program.

In the process of auditing and reviewing each function, R&QA defines and documents each element of a program so that its requirements are not subject to the vagaries of interpretation. Also, by establishing controls, the company is less likely to repeat errors since corrective action is recorded for future action.

The Quality Assurance staff provides consulting services to the rest of the company on quality matters. The technological aspects include fitness for use decisions, vendor selection, code interpretations,

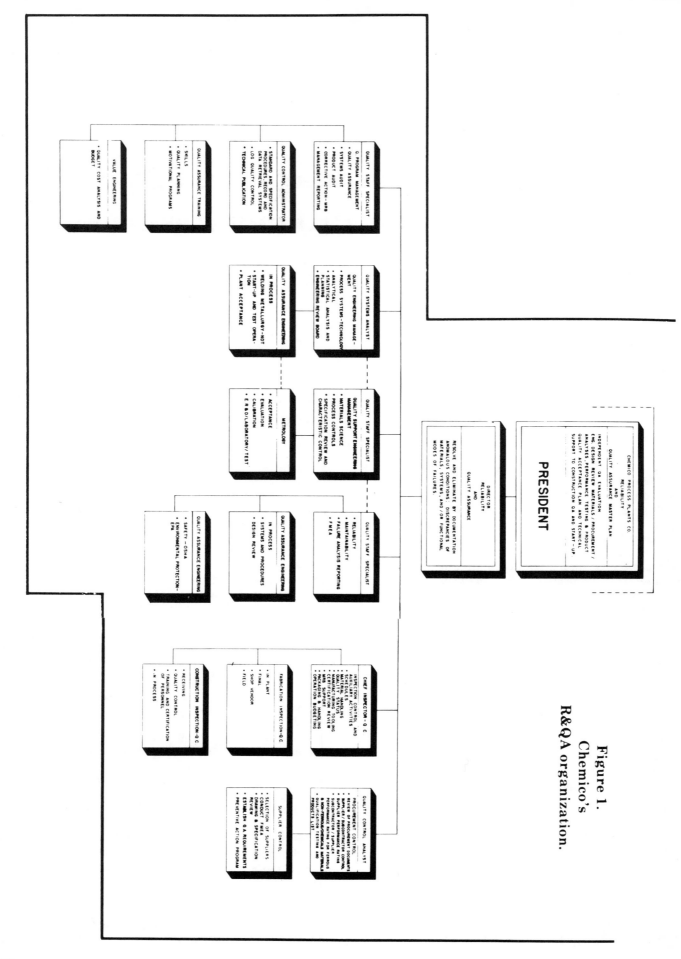

Figure 1. Chemico's R&QA organization.

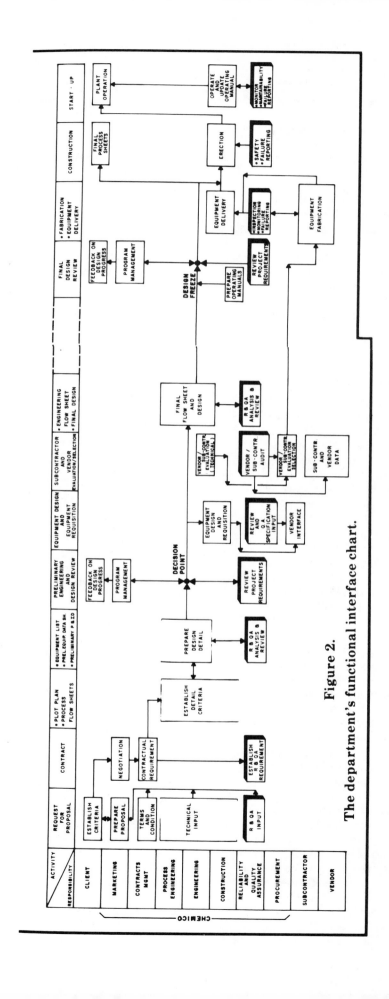

Figure 2. The department's functional interface chart.

29

welding procedure qualifications, and material and product specifications. On the corporate level, Quality Assurance provides inputs on the relation of quality to the business goals of the enterprise.

The functional organization of the R&QA group is shown in Figure 1. The department is headed by a director who reports to the president of the corporation, with the rest of the staff reporting to the director. This centralization of services and functions assures the optimum concentration of responsibilities and provides a single interface with other Chemico departments and outside organizations, as shown in Figure 2.

The company's policy objective is to deliver a quality product consistent with the requirements of a contract. To provide effective direction and measurement of quality activities, a program plan that encompasses all action from contract initiation through product delivery is prepared for each project. R&QA responsibilities to this plan are listed in Table 1.

Design and development controls

Compliance with quality criteria during the design and development phase is assured by an established, documented program providing for review by the project and process engineers, and by appropriate members of the General Engineering and R&QA Depts. The program manager provides general oversight for the design review activity and the project engineer chairs the meetings of the design review group. Quality checklists prepared by R&QA for use during the design review provide the basis for inspection, test, and quality of the article under review. The Quality Assurance representative ensures the inclusion of sufficient provision for determining and controlling the quality of the items throughout the manufacture, assembly, test, and end use. Figure 3 presents a typical flow diagram detailing the design review process.

In conjunction with design review, R&QA performs specific analyses and evaluations to provide further insight into the process design. The group provides support to the design and process engineers in the form of trade-off studies and analyses evaluating new processes and alternatively proposed systems from a reliability/quality point of view. Also, components and materials are evaluated for use in the various processes and systems. The objective of the evaluation is the selection of materials and components having an adequate reliability/quality background for the intended application. The failure history of the components and materials are investigated to ascertain their acceptability for use.

The historical data file compiled provides a useful source of information during the design phase. The availability of information on past performances, classified in a file system for ready accessibility, provides warnings to future projects of potential trouble spots, thus avoiding recurring mistakes and problems. The specific types of information recorded in the historical data file are:

1. Project related R&QA information such as plant failures, startup problems, deficiencies, trouble reports, maintenance problems, etc.
2. Vendor related R&QA information such as equipment failures, vendor quality ratings, shop visits, audit reports, etc.

The department is also responsible for monitoring the policies and performance of each participating supplier. It reviews purchasing documents with respect to their completeness in stating design, reliability, and quality requirements.

Procurement sources are selected and approved based upon the supplier's quality history records, and/or upon facility and quality system survey reports. The procurement documents are examined for the following items:

1. Contents
    a. Supplier quality programs
    b. Basic technical requirements
    c. Detailed quality
2. Raw materials used in purchased articles
3. Preservation, packaging, and shipping
4. Identification and data retrieval
5. Inspection and test records
6. Chemico's R&QA activity at source
7. Equipment records

R&QA ensures that internal policies, procedures, and instructions developed for control of the quality of all fabricated articles at a supplier or at the site is documented and maintained for effective operation and communication. All planning, fabrication, test, and inspection is based upon the latest documents to ensure that the completed article conforms to the applicable contract, drawings, and specifications.

In order to achieve process quality, uniformity, and accuracy, the department requires that fabricating and inspection personnel be trained and certified prior to job assignment and that recertification is performed periodically. In addition, cognizant facilities are regularly audited by Quality Assurance for conformance to controlling documents.

Inspection and tests

R&QA establishes and maintains a timely and effective inspection system covering the period from the procurement of raw materials through test of finished articles to provide for the discovery of defects or other unsatisfactory conditions at the earliest practical points.

Quality Assurance coordinates its activities with engineering and fabrication planning groups during all phases of fabrication to ensure the transmittal of quality information to all control points.

Inspection and test procedures are reviewed by Quality Assurance and used as applicable at each control point. The procedures state the criteria for acceptance and rejection; where applicable, workmanship standards are referenced to in-house as well as to industry standards.

Provisions are made to ensure that testing is done in accordance with test specifications and procedures including the accurate and complete recording of data and test results. Any nonconformance, rework, repair, or modification occurring during the test is documented.

Following the testing, Quality Assurance ensures that remedial or preventive action has been accomplished relative to nonconformances. Staff personnel verify that test results and reports are accurate, complete, and traceable to the tested articles.

A Material Review Board (MRB), set up by the department to examine nonconforming articles, consists of the cognizant design engineer and Quality Assurance representative. Decisions to accept such articles is by mutual agreement of these two people. MRB decisions are documented to show the details of the irregularity and the disposition of the item, be it rework, repair, use as is, or scrap.

Each nonconformance is investigated to determine its nature and cause. When action to correct any deficiencies and to minimize their recurrence is decided upon, the group ensures that the corrective action is taken, and that documentation of the incident is adequate. A file of MRB actions is maintained to aid in providing immediate information of any recurrence of the deficient condition.

Reliability and maintainability

If required, due to process complexity or new developmental technology, a reliability analysis for assessing reliability achievement can be developed and implemented in the early phases of a program. From these analyses, a logical and documented assessment of equipment/process risks and uncertainties can be derived. A reliability assessment analysis includes a system reliability block diagram for the entire process under consideration. Using this diagram, a Mean Time Between Failures (MTBF) estimate can be generated.

Failure Mode and Effect Analysis (FMEA) is usually performed at the principal component level for analyzing possible failure modes. Levels of criticality are ranked. The FMEA isolates and weighs critical failure causes into a quantitative index which permits a continuous monitoring of reliability. The findings are used for:

1. Identifying design deficiencies, limitations, uncertainties, and problem areas.
2. Evaluating designs and recommending design improvement.
3. Identifying critical failure modes that would negate the attainment of program objectives.
4. Determining the need for changes in the test program.
5. Assisting the generation of technical and operation manuals.

Table 1. R&QA objectives and functions.

Objectives	Functions
1. Conformance of parts, assemblies, and units to approve specifications.	1. Develop, coordinate, and/or review and approve major quality and reliability improvement programs.
2. Testing meets satisfactory performance criteria simulating operating conditions.	2. Analyze data from various sources and transmit significant information to other departments.
3. Documentation fully satisfies client's requirements.	3. Administer program for control of suppliers quality.
4. Elimination of discrepant fabrication through proper coordination.	4. Administer program for assuring satisfaction of client's contractual requirements including 1) review against referenced specifications to confirm compliance, and 2) surveillance of components in service.
5. Improvement of skills and attitudes through the establishment of training and motivational programs.	5. Implement controls necessary to provide and maintain required quality and reliability levels.
6. Improve quality through the lesson of experience.	6. Issue publications including quality/reliability procedures, specifications, manuals, and handbooks.
7. Establish quality criteria appropriate to the product requirements.	7. Audit quality and reliability performance, and maintain a system for corrective action and reporting to management.
	8. Administer material review and design review functions.

In conjunction with the diagram, MTBF estimates, FMEA, and equipment failure history, specific recommendations will evolve as to redesign requirements, redundancy implementation, and elimination of safety hazards.

R&QA ensures that maintainability and human factors are considered during the design phase and that the final design has the capability of being 1) maintained by the performance of specific checkout, repair, and maintenance techniques, and 2) fabricated, handled, maintained, and operated to high efficiency with minimum hazard to personnel and equipment. Limited life components are identified as possible candidates for scheduled maintenance. Efforts are focused on the reduction of human-induced failures that may occur during handling, maintenance, and operation.

Cost considerations are a factor in identifying the most effective means for maintaining the process operation. Trade-off studies determine the advantage of scheduled/unscheduled maintenance vs. active/standby redundancy from a cost effective standpoint.

Where the needs of a program dictate, a formal system of failure reporting, analysis, and corrective action can be established and maintained during all phases of a program. All failures or malfunctions are reported to the R&QA Department and appropriate failure analysis initiated where necessary. This analysis is reviewed by cognizant engineering groups and based on the analysis a course of corrective action is determined. The department insures the depth of the analysis is adequate, the results are acceptable, and the corrective action is implemented. Figure 4 depicts a typical failure reporting system.

The benefits

Since the recent inception of the R&QA program outlined in this article, the following benefits have been realized:

1. Design reviews have been instrumental in providing completeness of criteria such that vagueness or ambiguity are eliminated, thereby avoiding a possible compromise of intent or effectiveness of the objectives.

2. Trade-off studies performed by R&QA for comparing process alternatives have provided a different insight or basis into the merits of each alternative. It has challenged various conclusions by engineering not supported by adequate data. Potential problem areas or areas requiring further investigation were recognized before the company's resources were totally committed.

3. Poor or marginal suppliers were reduced by a method of rating supplier performance based upon previous inspection history and current quality surveys of the suppliers' facilities.

4. The R&QA review of all Chemico standards and specifications for currency and applicability to

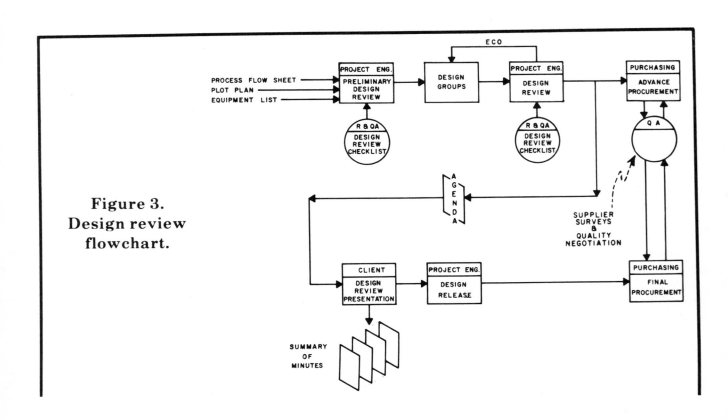

Figure 3. Design review flowchart.

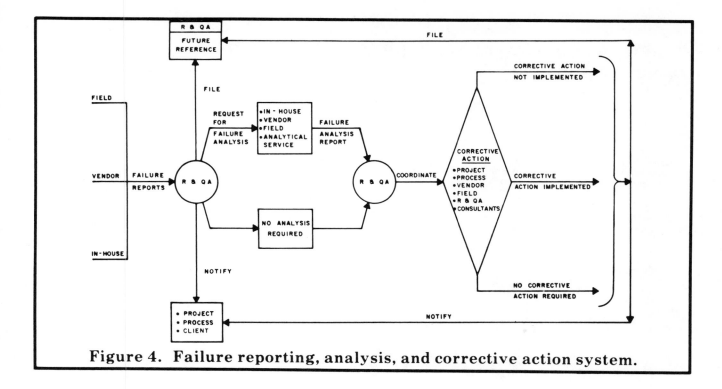

Figure 4. Failure reporting, analysis, and corrective action system.

industry standards (ASME, ASTM, API, etc.) has resulted in the detection of references to obsolete standards and incomplete and superfluous requirements in Chemico specifications.

5. Reference to the historical file has provided recognition of recurring problems, which left undetected, would have resulted in costly rework and schedule delays.

6. R&QA has developed and is in the process of implementing a documented formal program for training, upgrading and certifying inspectors in terms of their responsibility to code requirements.

7. It is instituting the operational controls to standardize the day to day activities throughout the company. This will eliminate the confusion and randomness of controls when transferring between different projects.

8. Because its boundaries extend across all projects, it has provided a link between the various projects for dissemination of data.

9. The institution of an R&QA discipline has been helpful in securing future contracts. Prospective clients have additional confidence in organizational capabilities since the company has taken steps to limit mistakes and improve its output.

The resultant cost savings and reduction in potential schedule delays are difficult to measure and evaluate. But, it is axiomatic, that a company that performs within cost and schedule restraints remains competitively strong.

Acknowledgments

I would like to thank E. L. Andersen and D. Livaccari for their help in preparing this article.

Bibara

Quality Management Through Quality Indicators: A New Approach

By Subhash C. Puri
Chief Statistician
Food Production and Marketing Branch
Agriculture Canada, Ottawa, Canada

John R. McWhinnie
Consultant
Organization for Economic Cooperation and Development
Paris, France

Total Quality Function

Total quality control is essentially a total management function. Even when the production/manufacturing process is basically a technical or engineering function, the ultimate responsibility for its effectiveness rests with the management. The total quality management function, being a many-faceted function, encompasses all activities and functions associated with the development of quality, viz., quality control, quality assurance, and quality reliability. Whether it be for the producer, the consumer, or the regulatory agency, the ultimate goal of any quality management function is to have at hand a product of superior quality i.e. a product that is satisfactory, reliable, and economical. This aim is generally achieved through and for the collective satisfaction of all the parties concerned with the quality of a product or process.

A Case Study

In this paper we shall consider the management of total quality as used by a governmental regulatory agency, and develop the concept of quality indicators as an aid to the management towards better organizing, planning, and decision making in their quality programs.

The function of a governmental regulatory agency is to assist the producer in realizing from the market a fair return for management, capital, and labour, and at the same time insure that a high quality product is being delivered to the consumer in accordance with the established standards and specifications. In order to carry out such a function, the regulatory agency develops a quality control program and establishes guidelines that include the following important features and procedures.

1. Develops national grades, standards, specifications, and tolerances compatible with the capacity and capability of the producer to produce a certain product and in liaison with the consumer preference and market demand.
2. Establishes regulations and legislation governing the control of quality of the product.
3. Establishes methods of inspection and/or monitoring of the product for compliance to the established standards.
4. Assists the producer in the smooth regulation and flow of his product, whether it be in terms of research, market information, management or any other related aspect.
5. Assists the consumer by rectifying any problems or complaint.
6. Develops methods (such as control charts or surveys) to keep track of the state of quality of the product in terms of revision of standards/specifications/tolerances and thereby the regulations.
7. Develops cost estimates, performance evaluation procedures to check the

validity, effectiveness, efficiency and economic aspects and usefulness of their quality program. This helps in further planning and organizing.

In general, a governmental regulatory agency assists in the continued development of a market-oriented total quality system designed to supply safe, high quality product at reasonable prices to consumer and to provide equitable return to producers, processors, and marketers.

The Concept of Indicators

The concept of indicators as a management tool for analyzing or monitoring a system made its debut in the field of economics (GNP is a classic example). Indicators have since been applied to many other fields such as health, the social sciences, or management performance to provide key information on a timely basis. An indicator can be a single measure or a combination of selected measures which provides simplified information concerning the state of affairs a system is in or will be in if one looks at trends over time. This is perhaps the first time that this concept is being proposed for the area of quality control. It is our contention that the total management function can be performed more efficiently and effectively through the use of quality indicators. In the case of the governmental regulatory agency, as in any quality control situation, there are many activities and phenomenon which are difficult to measure directly or totally. In such cases, a properly chosen indicator can reflect the status of the activity or phenomenon without monitoring a full-scale analysis.

Consider, for example, the inspection function which relies heavily on the amount of sampling and the selection of sampling plans. It is difficult to measure how much an inspector is capable of performing and what his efficiency level is. It would be helpful for management to have some readily available idea about the end results of inspection through quality indicators.

Types of Indicators

We have identified three types of indicators; result indicators, work indicators, and environment indicators. In order to search for the right kind of indicators, it is essential to identify activities and functions that display and explain performance; i.e. effectiveness and efficiency whether it be due to external factors or internal problems. One would also need the information that helps in managing that performance, for example, statistical information, intermediate performance data, reliability analysis data, cost/analysis studies, record of inspection, etc.

The result indicators can be found at the level of the program or major activity or at the operational level; for example, a statement of program objectives is found in the estimate or terms of reference for the responsibility centre. They should be identified with responsibility centres and should be consistent with agreed objectives, programs or operation. The work indicators can be found in management reports, statistical summaries, operating manuals, etc. They must describe work itself and not a decision arrived at as a result of the work. They must also be quantifiable and affecting the resource utilization directly. The environment indicators determine the desirable level of quality, the demand factor and all other factors that affect the consumer of the goods and services (outputs). The behavior of these factors as determinants must be forecastable and there must be a relationship between the selected environmental factors and the level of output.

Methodological Considerations

Selection of indicators is not a task to be taken lightly and should be approached with caution. There are several important characteristics of indicators which must be tested and proven before management can make use of them with confidence.

First of all an indicator must be valid - in other words, any change in the phenomenon which it represents should be reflected by a corresponding change in the indicator. An indicator should also be reliable - that is, it should give consistent results for the same phenomen. One must also take into account any possible seasonal, geographic or cultural biases which might affect the performance of an indicator. Any indicator should be rigorously tested and maintained in concert with conventional management information systems before being put forward on its own.

Once a series of valid, reliable and culture free indicators has been established, one might want to consider the incorporation of some or all of them into a global indicator, often called an index. The various components could be weighted according to their relative importance in the total quality management function such that a single monthly index could alert management to the status of their total quality program.

Application

In a simple situation of a governmental regulatory agency, it is not only profitable to develop the management program based on indicators but it makes the program much more easy to operate and more effective. For such a model some examples of important indicators of quality can be applied on the following:

1. Inspection Records:

    Inspection activity is one such function that can provide an indication to many aspects of quality. For example, the inspection records can show the state of quality and trends of quality of a product on the whole or for each producer separately. It can also indicate the efficiency of an inspector. All these records can be kept through control chart method and analyzed statistically. Thus properly kept inspection records will provide a good set of indicators, and the management can plan and organize the whole program accordingly.

2. Cost Figures:

The cost figures for the whole program can indicate whether it is worth running in terms of money. In particular the costs incurred on the inspection can provide indicators whether the level and amount of inspection is justifiable or not. Acceptance sampling plans can be found to suit the reduced inspection if necessary. This would reduce costs while maintaining the level of efficiency.

3. Specifications:

The patterns of average outgoing quality of the product would indicate whether a change is needed in the AQL (acceptable quality level) or specifications. Standards, specifications, or tolerances should be kept compatible with the actual quality of the product.

4. Consumer Reports:

Consumer preference, reports, or complaints should serve as an indicator to the overall state of quality of the product.

5. Reversals:

A reversal occurs when a product at the shipping point is inspected, under a certain sampling plan, approved for sale to the buyer but is rejected on a second inspection by the buyer at the point of destination. The product is sent back to the producer for rectification. The number of reversals for a product or producer should indicate the state of quality for the product or producer.

While other areas can be selected, it is wise to keep the number of indicators to a minimum such that key development in the overall process can be spotted on a timely basis.

Conclusion

The point that we wish to emphasize in this paper is that the usefulness of indicators as a management tool can be exercised in the field of quality control. Planning with the help of indicators can bring better results than achieved through a cumbersome detailed management information system. A useful by-product of using this technique is that it will improve communication among various components within the program, an element which is generally lacking in many situations. Lack of coordination is usually the cause of failures of programs. Since the establishment of indicators requires information from all sources of the quality implementations, there would be an increased participation and better communication. The lines of communication can be extended outside the program also, i.e. to the producer as well as the consumer. This collective participation will provide better feedback, improve upon the quality of indicators, and bring about a collective satisfaction thereby resulting in a superior product.

The identification and development of a set of precise, valid, and reliable indicators for the total quality management program of a particular governmental regulatory agency will be reported later in another proceedings.

16th QUALITY ASSURANCE CONFERENCE, NEW COLLEGE, OXFORD, SEPTEMBER 1976

Quality Assurance and Other Departments

by David J. Murray

This paper, whilst insisting that quality assurance is a specialized function to be undertaken by people with particular knowledge and skills, reviews the breadth of the activities. It is an operation carried out on many fronts. Departments with which an effective quality manager must work constructively are listed and some of the areas for liaison are considered. Finally, it is emphasized that quality does not automatically arise from a theoretically correct organization structure but from appropriate and efficient interations between people.

Introduction

If Quality Assurance staff are to play a full and effective role in an organization they must establish good relations with many other departments. The obvious areas are Manufacturing, Design, and Research & Development. The relations between these departments and Quality Assurance have often been discussed by others and this paper will not attempt to go into detail about them. However, there are many other parts of the organization whose activities can either significantly affect the quality of products or provide information about it.

Before proceeding further I will define what I mean in this paper by a Quality Assurance Department. It should not be confused with an inspection section which, although extremely important, performs only part of the overall quality function. Rather, I refer to that department which is responsible for the establishment of company-wide policies, procedures, and standards related to quality, within which other departments, whether or not their prime function is quality control, will operate.

There is often a debate as to where in the organization a Quality Assurance Department ought to be placed. With the development of so many managerial specialties during recent decades there has been a tendency for each emerging 'profession' to claim that it has a special right to report directly to the managing director or chairman of the company. Quality assurance specialists have not been totally preserved from this temptation. Many seem to assume that they cannot possibly operate effectively without having the head of their department in the Board Room, clearly distinguished from manufacturing and other technical directors and managers. In some organizations this may be the best way of operating. However, it would be overstretching a point to assert that there is only one possible correct reporting position.

Many words have been written in recent years on the question of divorcing responsibility for quality control from the responsibility for manufacturing. Again, in some firms this may be essential, but it is not necessarily a criminal offence to have the Quality Assurance manager reporting to the production director. These matters must be decided after considerable thought at Board level, which will take into account the history of the organization, the present stage of its development, the personalities in various roles, the technology and products, and many other issues.

Whatever Quality Assurance reports within the structure, however, it must be clearly recognized as a distinct specialty and not simply an extension of the Inspection Department, or (just as bad) a bit of unfortunate bureaucratic paperwork tagged on to the end of design or research activities. The larger issue of the emergence of quality assurance as a profession is in a state of flux at the present time,[1] but whether it is to be classed as a profession or not it must still be seen as a separate specialty requiring people of different type and calibre from those needed for other technical and managerial activities within a company.

A discussion of some other aspects of relations between quality assurance and manufacturing can be found in my paper given to the European Conference on Quality Control in Copenhagen.[2]

Departments to which
Quality Assurance must relate

Sales/Marketing.

The degree of involvement of Quality Assurance with Sales and Marketing will depend, among other things, on the type of products being made and the markets to which they are being sold. For example, industrial products which are tailor-made for specific customers may well demand a close liaison between Sales/Marketing staff and Quality Assurance. Where standard products are being made in bulk and sold to a wide range of customers the primary market-technical interface may be between Sales/Marketing and Design/Development departments. However, as Quality Assurance ought to be involved in specification development, there clearly should be contact.

Specifications must be written from the user's point of view. If one accepts that quality is 'fitness for purpose', it is vital that the writers of specifications in all parts of the organization should be aware of that purpose. There is no point in having the most comprehensive technical test specification if it does not relate to the requirements of the market. One of the prime functions of a Marketing department is to establish what the customer requires.

Sales literature must often include usage or operating instructions. A package of paint, for example, must carry instructions for use. The end objective is not the liquid in the tin but a decorative or protective film on a door or a wall; the wording of literature is therefore vital. The image of one's product can be determined as much by the education and skill of the user as by the product itself. Similarly with many consumer-durable products emphasis is increasingly being placed on total life-cycle quality so that it is necessary for careful attention to be given to operating and servicing instructions. This is also especially relevant to capital plant for industry as total life-cycle costing becomes more common.

The sales force, as the nearest point of contact to customers, will be the main collector of complaints and rumblings of dissatisfaction. The Quality Assurance Department should liaise closely with the sales office and establish a clearly understood and efficiently operated complaint-handling routine which, as well as satisfying the customers, will produce management data on frequencies and trends.

Planning.

The role of operational planning departments varies considerably from company to company. In this paper I will just refer to four aspects of their work.

They are frequently concerned with new product planning, and close liaison is necessary to ensure that quality is considered and managed into all stages of a new product introduction. Usually, Inspection Departments near the shop floor will succeed in putting in appropriate testing arrangements whether Quality Assurance has been involved at an early stage or not, but I am a firm believer in the concept of prevention rather than cure, and hold that the close involvement of Quality Assurance in new developments can prevent difficulties at a later stage - if only by asking awkward questions early in the project.

Levels of stocks are frequently planned by central departments; Quality Assurance people will want to watch stockbuilding activity very closely especially when products have finite lifetimes. From time to time it is necessary to transfer the manufacturing of products from one plant (or one part of a plant) to another; it is often beneficial to have a register of approved plant areas in which each product can be satisfactorily made and where the necessary expertise and peripheral equipment (say for testing) is available. Quality Assurance people may also wish to challenge planned rates of production, especially if there is likely to be a contrary correlation between speed and quality.

Purchasing.

Every organization could benefit from having a general policy on materials purchasing and quality. This could cover such items as the approval of sources, the procedure for arriving at agreed specifications and sampling schemes, arrangements for handling noncomforming deliveries, evaluation of supplier performance, and other topics. It will be necessary to obtain deep involvement of the senior manager responsible for purchasing, and also the manager in charge of routine buying, during the formulation of the general policy. However, it will be even more vital to have their co-operation during the implementation stages, and in regular audits.

Warehousing/Distribution.

Many products can be seriously damaged by storage under wrong conditions. For example, in the paint industry emulsion paints must under no circumstances be stored during cold weather in warehouses without adequate heating; similarly, most gelled gloss paints should not be exposed to high temperatures either close to heaters or in the full glare of the sun through glass. Methods of transport can also have serious effects. Such factors should also be considered in the storage and transport of raw materials and intermediates.

Stock rotation is a further quality determinant; many products deteriorate with age, apart from difficulties due to obsolescence following design changes. In our company, therefore, one of the principal duties of the Quality Audit Department is to examine stock regularly in our main factory warehouses and in the many regional branches distributing our products throughout the United Kingdom.

Legal.

As in all sections of industry the law is now becoming more important to the Quality Manager. Legislation covers labelling, performance levels of many products, weights and measures, transport, packaging, and hazardous materials. These developments come both from the United Kingdom government and from the E.E.C., as well as from an increasing number of international, national, and other agreements, and regulations which, whilst not strictly statutory, have in effect the force of law in that one has no alternative but to comply with them. Consumer legislation is continually increasing, as reflected in the fact that the main stream of papers at the recent Copenhagen conference of the E.O.Q.C. was largely devoted to developments in product liability and similar legislation related to a social conscience. The United Kingdom government is currently considering developments in this field and has recently circulated a consultative document. All of this means that the Quality Manager must be able to go to his legal department for advice on interpretation of the new legislation. To a large extent he may (along with production management) be held responsible for ensuring that products consistently comply with requirements.

Accounting.

The primary association between Quality Assurance and the accounts department, apart from the normal control of departmental operating budgets, is in the field of quality costs. I will not go into detail on this subject because the matter has been extremely well dealt with in a recent issue of Quality Assurance under the general title "Quality IS Profit". Suffice it to say that many companies have found that a clear identification of the elements of quality costs has enabled them to make considerable savings, and therefore profit improvement.

Plant development/Engineering.

Virtually all companies have some form of engineering or maintenance department, but many also have separate units within the organization dealing with important plant and process developments. Quality Assurance may gain considerably by establishing with such departments the factors which need to be carefully considered during design and construction because of their possible impact on quality. These could include such items as selection of constructional materials and measuring equipment, possibilities for contamination and errors, facilities for cleaning, control systems and ergonomics, and plant flexibility. Clearly the factors requiring consideration will vary from industry to industry but it is far better that Quality Assurance people should be involved in the early design stages than that they should be called on to develop compensatory inspection schemes at various points in the process when it is too late to alter the basic plant design.

Systems development.

By systems development I refer to the development primarily of commercial systems. In the more advanced companies this may involve the use of extremely complex networks of on-line computer video-terminals recording everything that takes place in the movement of materials and money from receipt of an order to payment by the customer; indeed, it may go further and extend to the production of detailed accounts and management information, including forecasts of future demand based on a combination of historical data collected by the system itself and factors put in at the time of requesting the predictions. When one begins to think carefully about these types of systems one realizes just how much technical management information can often be obtained with a comparatively small amount of additional effort. Parts-lists and product formulations relating to every product will probably be on computer file. This gives an opportunity to carry out various analyses of components by computer, as well as to compare actual production batches with the theoretical item-contents in the master files. It may even be possible to enter test information along with data regarding quantities produced; this can then be analyzed in various ways, for example producing routine control reports ranging from minutely detailed reports to summaries for senior management.

Packaging.

In recent years more emphasis has come to be placed on packaging, and it is now often realized that the quality of a product when it reaches the customer may be affected by the package in which it is contained. As a result in our company the technology of packaging is studied within one of the Quality departments. Not only must we ensure that the appearance and strength of the package do not deteriorate unacceptably on storage and handling, but interactions with the product have to be studied. There may even be interactions between various elements in a packaging system.

Owing to developments in consumer legislation greater demands for product recall are likely in coming years; it is therefore necessary to have batch codes on packages of many types which in the past have not carried such marking.

Personnel/Training.

Quality is determined to a tremendous degree by the calibre of the personnel employed in the organization. Whilst it cannot be expected that the Quality Assurance Department be involved in the actual recruitment of people, except for its own activites, it ought to have something to say to the Personnel Department regarding the types of people suitable for some types of work. The Quality Assurance Manager may also be able to make a contribution to discussions on salary and other payment systems; these may strongly influence the quality of people available. The effects of bonus schemes may be either positive or negative as far as quality is concerned. Job-grading systems may also be significant in determining whether the correct type of people are obtainable for the work to be carried out.

The Quality Assurance Manager may be deeply involved in the formulation of training programmes - induction training, basic technical training, specialized technical training, and detailed job training at plant level. It is not likely, nor advisable, that he should himself conduct much of this training, but he should be involved in determing its content.

Safety and Environmental.

There are many points of contact between safety, the environment, and quality. For example, Quality departments will probably hold large volumes of technical information relating to material hazards. This is a prime concern of safety departments. As the amount of monitoring of effluent disposal into rivers, lakes, the atmosphere, etc., increases the Quality departments will in many companies be the only people qualified to measure such parameters in a systematic and scientific way. Whether the main responsibility for these matters ought to be in the hands of the Quality Manager or whether he ought simply to provide a service to another department is another of the many points which must be solved by each company in a way suited to its own situation.

Conclusion

In the main section of this paper I have attempted to give a quick survey of many of the reasons why a Quality Assurance department needs to have a broad spectrum of contacts within the organization. This is without discussing such subjects as service monitoring and quality control of clerical work. In thinking through the issues of suitable personnel for a Quality Assurance department can sometimes be difficult. Its members need to have a broad view of the operations of the organization. The senior manager should ideally have managerial experience or postgraduate education of a type which will have given him a 'wide-angle view' of the business. This, incidentally, is one of the reasons why Research & Development personnel frequently find it difficult to make the transition to Quality Assurance, and the re-orientation period can be lengthy. The specialist nature of their training has tended to cause most design and development staff to have a somewhat narrow view of industrial life. Many attempts are being made to improve this area of our educational system and we can look for a broader understanding among the next generation of scientists and technologists.

It has already been said in the first paragraph that there is no one single organizational pattern for quality assurance. I am a firm believer in this statement. To expect the establishment of anyone's favorite organizational structure to produce the results is naively optimistic, disregarding the imperfections of human nature. Whatever the structure and theory, the necessary links must be built up between real living people. A Quality Assurance manager must expend a considerable amount of effort in developing relations and establishing working arrangements with and between managers and staff of departments which affect quality. Only in this way can his objectives be achieved.

This paper does not deal in detail with the means of achieving working relations, but simply highlights the large numbers which are needed. Methods will vary considerably from one organization to another, and range from the very formal to the personal and informal.

Acknowledgements

Thanks are due to Mr. W.B. Giles, Managing Director of Crown Decorative Products Ltd., Paint Operations, for permission to publish this paper. Many of the ideas contained in it have been developed from work carried out within Crown, but it must be stated that the opinions expressed are those of the author and do not necessarily reflect the policy of the company.

References

1 Juran, J.M. "Emerging professionalism in quality assurance." Quality Assurance 1976 Vol. 2, No. 3 pp. 71-77. The Second John Loxham Lecture.

2 Murray, D.J. Paper delivered at E.O.Q.C. Conference (Stream 3) at Copenhagen, June 1976.

David Murray is a consultant with HAY Management Consultants, working from their office in Manchester, England. A chemist by initial training, his industrial experience includes research and development, quality assurance and production management. In the mid-1970's he was founder-secretary of the Institute of Quality Assurance study group on quality in process industries and in 1977 received the Queen Elizabeth II Silver Jubilee Award for contributions to the profession of quality assurance.

Many companies, particularly those producing critical products, have had quality assurance programs for years. There are, however, a multitude of firms still struggling with QA design and implementation. The following article discusses a complete QA program by outlining the basic elements of (1) administration, (2) procedure and (3) appraisal.

DEVELOPING A QUALITY ASSURANCE PROGRAM

by **John H den Boer**
Director, Quality Assurance
& Metallurgy
Mill Products Div
Reynolds Metals Co
Richmond, VA

The origin of quality assurance responsibility in a company is its published policy on quality. If a quality policy does not exist, it must be established. Once established, procedures and instructions can be developed to support this policy.

The company's organization chain should be such that QA personnel do not report directly to lower line functions responsible for production, **Figure 1.** At the manufacturing level, the QA manager should report either directly to the plant manager or to him through the technical functions office. This prevents conflicts of interest.

In multi-plant organizations, it is common for this manager to report administratively to the plant manager and functionally to a division or corporate QA manager. No matter how the QA departments are organized, it is important to have a clear statement of the organizational structure and of the responsibilities associated with each of the positions in the structure.

Records should be formatted to provide clear information. Each type of record should have a unique title and clearly identify the data recorded.

On the other hand, reports need not all be formal, but it is a good idea to standardize the format. This helps a reader to make valid comparisons. Reports should also be titled and dated. Their distribution should be limited to those who have responsibility for taking corrective action, either directly on processes affecting the data (feed back) or indirectly on other processes that would be affected (feed forward).

Retention of records and reports should be considered. There is no set length of time a record should be retained; however, several things need to be considered.

• Review of data. There may be a need to review data when changes have

occurred for unknown reasons. Generally, a retention of one year is adequate.
- Confirmation of data. There may be a need to confirm data relating to specifications or production. The statute of limitations in most states is seven years from the time of occurrence of an accident. Judgment will have to be made as to the maximum life of the product. The retention time for important records should be the sum of the maximum life and seven years. These are considered primary product liability records.
- Back-up data. Other substantiating data in records, defined as secondary product liability records, may be retained for a shorter period.

The primary cost of record retention is collecting and placing the data in storage. When storing data it is important to consider frequency of retrieval. For example, data used to review product changes would probably be needed more often than product liability records. Once in storage, the cost for retention is relatively small, so retention should be for as long as possible, considering the available space.

Record audit trails should be prepared to assist in tracing records. Traceability should be both forwards, from production to the customer, and backwards. Information needs to be recorded accurately and consistently on all related records to avoid loss of traceability.

Consideration should be given to the design and distribution of a QA manual. Users should not have to wade through a thick manual each time they have to refer to another page. They won't do it. The information should be on as few pages as possible. Distribution should be based on the need to know. The manual can be divided into sections applicable to work areas, products etc, so that only the information necessary to a particular interest group is distributed. Revisions should be distributed as promptly as possible.

QA procedure

QA procedure consists of (1) design and engineering, (2) operating practice and (3) quality control procedures. Design and engineering involves planning and preliminary work such as product design and review, prototype testing etc.

Product design and review requires consideration of reliability and liability. There should be formal reviews to cover both aspects, and these reviews should be documented. During prototype testing, data is accumulated to prove whether or not a product can be expected to meet reliability requirements. Design and prototype testing will provide information necessary to identify future quality control instructions.

Release procedures must guarantee that no product is marketed until all design and engineering requirements have been properly met.

Operating practices include technical practices and procedures, quality control instructions and process procedures for both internal operations and vendors. Examples of technical practices and procedures are the instructions which must be followed to produce a product to specification.

Quality control instructions are additional instructions oriented toward other aspects of a product's specifications. They are generally of a technical nature such as dimensions or weight.

Process procedures are the more general methods used to produce the product. These controls are necessary to assure consistency of the final product. Control on variability of purchased components may be very important to a product's consistency. In this instance, consider controlling the vendor's procedures, watching for any changes, and evaluating their effects.

QC procedures differ from QC instructions in that there is an exerted control causing that product or process to conform to a specific requirement. To be properly designed, the QC procedures must address the following items: sampling procedures, testing procedures, acceptance limits, indication of inspection status, control of nonconforming material and corrective action.

Sampling procedures include number of samples, frequency of sampling, methods for selecting samples and responsibility for taking samples.

Testing procedures should be sufficiently detailed to cause each person following the procedure to produce essentially equivalent results. If there is a possibility that two people could arrive at significantly different results because of differences in techniques, then the procedure should be defined more sharply. In all cases the equipment or instrumentation used, as well as the standards used, should be identified. The persons responsible for making the tests also need to be

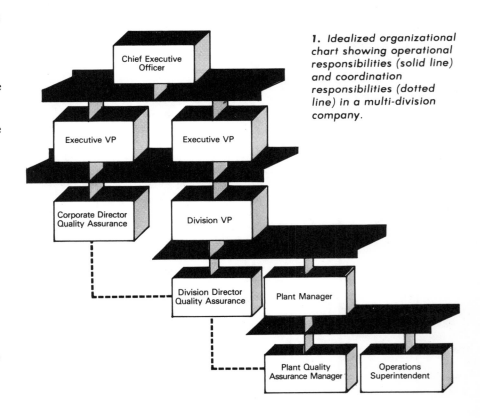

1. Idealized organizational chart showing operational responsibilities (solid line) and coordination responsibilities (dotted line) in a multi-division company.

identified. Methods for recording data and the types of reports to be prepared should be described.

Acceptance limits are values describing the range of tests or measurements considered acceptable. Values do not need to be numerical, but do need to be in terms of the measurement made and unambiguous. Floating limits must not be used.

Indication of inspection status occurs after the tests are completed and a determination has been made whether the material or process tested is conforming or nonconforming. This information needs to be announced in an understandable form to others associated with the process or product. Is it usable or not? Can the process continue or not? Does anything need to be changed? This announcement can take the form of tagging or marking of the material, movement of material to specific locations or issuance of reports.

It is necessary, once it is established that a material is nonconforming, to see that it does not continue to flow with the conforming material. The simplest way this can be accomplished is by placing it in the scrap can. But even this can lead to trouble if the scrap can is not properly identified and is similar in design to the tote boxes.

When there is rework or repair to be done in order to bring it back into the flow of conforming material, nonconforming material should be separated by placing it in a different area. It is best to have a specific area assigned for this purpose. The area can also be used to hold the material until final disposition can be made.

Tagging can be used on material at any stage—incoming, in-process or finished. The problem with tagging, especially if the material is not removed from the main line flow, is that the tag may be overlooked. This makes it imperative to identify all persons responsible for controlling nonconforming material.

Corrective action closes the QC procedure loop. It is probably the most important element from the standpoint of making the control effective, the simplest one to describe and the most difficult to implement. In the development of corrective action procedures, the primary objective is to identify the person(s) responsible. This is necessary to identify the basis for determining if any action is necessary. It is not realistic to make adjustments every time a result is out of specification.

QA appraisal

Appraisal of a QA program is generally the least understood and least developed part of the program. Areas of concern are audits, reviews and analysis. There are many forms and purposes of audits, but all of them have one thing in common—the collection of information. This information may be collected to determine whether certain stipulated procedures are being carried out as planned, the effectiveness of procedures or working personnel, or any other subject of interest to the auditor.

There can be various degrees of formality to the audit. But there must be a plan and it must be done with sufficient regularity and scope to identify problem areas. The plan serves to guide the auditor in the development of the information he collects.

An audit plan is usually used more than once. It assists in making the results of one audit comparable to those of other audits made to the same plan. When two or more people are making audits separately to the same plan, they should have a basic agreement on the intent and the general approach. For formal audits, there should be an entrance interview and an exit interview followed by a formal report. This report includes the agreement made for the corrective action plan. For informal audits, it is sufficient that a report be issued attesting that the audit was made and identifying the findings and corrective action. In all cases, there needs to be follow-up on the corrective action to assure that deficient conditions are corrected.

Reviews differ from audits in that they are not structured. Test or inspection data should be reviewed with the intent of determining whether or not corrective action has been effective. If it hasn't been, there could be a need for training or a major change in the procedures to improve controls. This review can be made by examining averages, standard deviations, ranges, failure modes and, on statistical charts, the frequency of out-of-control conditions.

Review of complaints, rejections or any other comments or activities that reflect on the quality of the product sold can indicate whether normal procedures are satisfactory, too stringent or not stringent enough.

Once a year, a review should be made of quality costs. For each review, there should be a report that includes recommendations on the action that management should take toward improving quality or productivity, or for reducing the cost of quality.

Audits and reviews report on known quantities, prescribed procedures and recorded results. Analysis relates to what is unknown but should be known. It is the intent of analysis to (1) discover hidden dangers that exist in the procedures for manufacturing a product that could result in quality problems in the final products, and (2) discover hidden opportunities for increasing productivity or general quality level of the products produced.

When analyzing product liability, ask the questions: "What are the worst things that could happen? Do we have those in control or under observation?"Various inductive and deductive tools, such as fault tree analysis, could be used to identify factors that could cause problems. Determination could then be made as to the effectiveness of the existing controls or the need for additional controls. ■

About the author

John den Boer received a BS in metallurgical engineering from Missouri School of Mines in 1948 and an MS from Vanderbilt University in 1970 with a major in engineering management. He has been with Reynolds Metals Co for 31 yr working as a metallurgist or in quality assurance. This article is based on his presentation at the 34th annual technical conference of the American Society for Quality Control held in Atlanta, GA.

CHAPTER 3

METHODS OF QUALITY CONTROL

Methods of Quality Control

by Hans J. Bajaria
Associate Professor of Mechanical Engineering
Lawrence Institute of Technology
Southfield, Michigan

Quality control is a science that uses quantitative and qualitative methods to make decisions that affect the quality of manufactured products. These methods simply provide the tools to analyze the information at hand in order to make accurate decisions. For example, one may judge the acceptability of a batch of goods by using one of the quantitative methods to measure a representative sample taken from the batch. Or one can use Pareto analysis to determine which factors are causing the most problems in order to allocate his resources judiciously. Some of the quantitative and qualitative methods are discussed below.

A. Quantitative (Statistical) Methods

Quantitative methods include that group of techniques which allow engineers to make decisions based on numbers. These methods, in many instances, require statistical treatment of the data. Some of the methods fall in semi-quantitative category in the sense that the numbers involved for statistical analysis are the conversions of subjective feelings rather than objective measurements. However, they still offer advantage over any other method, since they reflect the best of engineering experience. Some of the quantitative methods are described below.

1. Statistical Quality Control

   The word "Statistical" means "having to do with numbers", or more particularly with "drawing conclusions from numbers".

   The word "Quality" means much more than the goodness or badness of a product. It refers to the qualities or characteristics of the thing or process being studied.

   The word "Control" means "to keep something within boundaries", or "to make something behave the way we want it to behave".

Taken together, and applied to a manufacturing operation, the words Statistical Quality Control mean this:

STATISTICAL	•	With the help of numbers, or data
QUALITY	•	We study the characteristics of our process
CONTROL	•	In order to make it behave the way we want it to behave

The term "process" as used in the previous statement, is capable of assuming many different meanings.

A "process" is any set of conditions, or set of causes, which work together to produce a given result. In a manufacturing plant we usually think first of a series of fabricating and assembling operations which result, for example, in the production of cable, electron tubes, relays, switchboards and other apparatus or equipment.

However, since the word "process" means merely a system of causes, a process may be far simpler than the ones mentioned above, or it may be far more complex. In statistical quality control the "process" we choose to study may be:

* a single machine, or a single fixture or element of a machine.

* a single human being, or a single motion performed by a human being.

* a piece of test equipment.

* a method of measurement or gaging.

* a method of assembly.

* the act of typing (or performing any clerical operation).

* a group of many machines turning out different or similar pieces of product.

* a group of many human beings (for example, a pay group or a Shop).

* a combination of human beings, machines, materials, methods, pieces of equipment, etc. For example, the procedures required to manufacture switches, rubber or wire.

* a method of processing, such as chemical treatment or plating.

* a mental activity, such as visual checking or making calculations.

* intangible human elements, such as attitudes, motives and skills.

* the whole mass of causes which result in a Guided Missile System (for example), including thousands of resistors, capacitors, tubes and other components furnished by hundreds or thousands of different suppliers.

* anything else which results in a series of numbers and is connected with unsolved problems.

In its narrowest sense the term "process" refers to the operation of a single cause. In its broadest sense it may refer to the operation of a very complicated "cause system". This is why it is possible to make "process capability studies" in connection with practically any type of Engineering, Operating, Inspection or Management problem; including such problems as overall merchandise losses, the overall cost of maintenance or plant-wide inspection ratios.

2. Design of Experiments

This group of methods allow the study of factors and their different levels as to their effects on the yield. It is possible to quantify individual effects of each factor and their interactions. The engineers can direct their courses of action based on these results.

3. Hypothesis Testing

This group of methods is a special group of design experiments. This group allows a study comparison of product characteristics, such as average performance, consistency, concentration, predictability, etc., between two products.

4. Correlation and Regression Analysis

This group of methods is yet another group of design of experiments. This group allows the study of relationships between two or more variables of interest in any given design situation. Such relationships are utilized for the best economic advantage in a given design or production situation.

5. Contingency Tables

This collection of methods allows tabulation of data when engineers want to investigate whether two or more variables have any meaningful relationship among them. If the conclusions from the analysis indicate that there may be a relationship, exact relationship can further be investigated by design experiments.

6. Summary of Data to Direct Future Design Actions

   Massive data can be summarized into single numbers called statistics. A study of pertinent statistics under any given situation can produce meaningful and economical future design actions.

   For example,

   study of "average" statistic indicates most likely performance of the product

   study of "standard deviation" indicates consistency of the product

   study of "skewness" indicates concentration of the product life toward the higher or lower possible values

   study of "kurtosis" indicates predictability of the product with respect to failure, i.e., a product may have fewer weak links and, therefore, it is more predictable or it may have more weak links and in that case it is less predictable.

   Many more quantitative methods exist to study any design or production situation of interest. If they are properly planned, they produce the needed answers most economically. Generally, a help from a specialist is required for economical designs of the tests.

B. Qualitative Methods

   Qualitative methods simply include either tabulating in column or row or matrix form, the information of interest in any given analysis case, and arriving at conclusions after looking at the information in that form. Some methods include representation of information in the form of diagrams rather than columns and rows.

   Many such methods have become popular because they offer a particular advantage over any other decision-making methods and also require no mathematical rigor.

   Examples

   1. Cause and Effect Diagrams (Ishikawa Diagrams)

      The technique consists of defining an occurence (effect) and then reducing it to its contributing factors (causes). The relationships between all the contributing factors are illustrated on a structure Ishikawa calls a "fishbone", Figure 12.

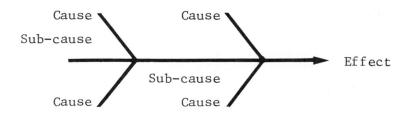

Figure 12 - Cause and Effect Diagram

The principle factors, or causes, are listed first and then reduced to their sub-causes, sub-sub-causes, etc. This process is continued until all conceivable causes have been listed.

The factors are then critically analyzed in light of their probable contribution to the effect. The factors selected as most likely causes of the effect are then subjected to experimentation to determine the validity of their selection. This method of selection and analysis is repeated until the true causes of the effect are identified.

2. Pareto Analysis

The technique of arranging data according to priority or importance and tying it to a problem-solving network is called Pareto analysis.

The technique identifies "vital few" high-priority items from "trivial many" in the problem-solving process. (Figure 13)

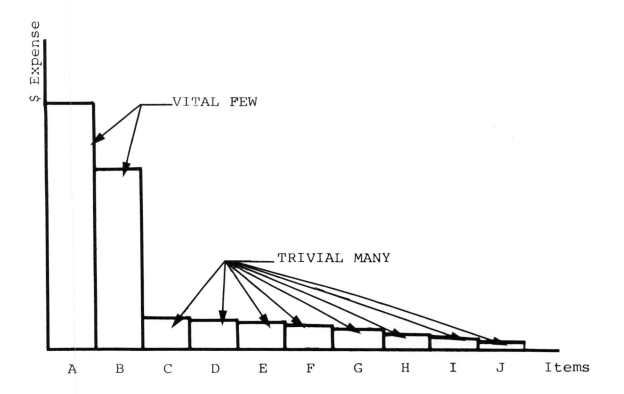

Figure 13 - Pareto Diagram

3. Kepner-Tregoe Problem-Solving Methods

   These methods, developed in the 1950's, break down problem-solving methods in three categories: (1) Problem analysis, (2) Decision analysis, and (3) Potential problem analysis. Once the problem category is identified, then action sequence corresponding to that category guides the problem solver to a solution.

CHAPTER 4

QUALITY ASSURANCE AUDITING

QA Auditing: A Managerial Approach

Robert E. Dellon
Assistant Director
Management Services Group
Washington Public Power Supply System
Richland, Washington 99352

INTRODUCTION

It is now more than seven years since Appendix B to 10CFR50 was formally issued as federal law by the then AEC. Since this regulatory requirement was first advanced, company after company has scrambled to develop and implement QA auditing programs in response to Criterion 18 of Appendix B.

Now that the nuclear industry has had significant years of experience with the QA auditing requirements of Appendix B, and experience with subsequent audit related ANSI standards, both draft and approved, I believe there is a need to put the managerial value of such efforts into perspective. Since there are sufficient standards, textbooks, handbooks, training courses and procedures available on what a substantive QA audit program must comprise to be effective, I will not dwell on any of these areas other than to state that planning is pivotal to the effort. Logical, realistic, programmatic planning must be used throughout the entire process, whether it be in the formulation of an auditing philosophy, policy, procedure or plan, or in the budgeting, staffing, training, or implementation of the audit function.

As you will quickly realize, I am a firm proponent of QA auditing. When executed with the strength of knowledge derived from training, experience, preplanning and professionalism, a comprehensive QA audit program can be an invaluable asset to a company.

FUNDAMENTAL INGREDIENTS

What are the fundamental ingredients of an effective QA audit program? As is true with any highly visible effort, the support of top management is basic - without it, for whatever reason, the program is rendered inadequate, a facade and a detriment to other QA activities.

Recognizing the need for management support, how does one gain this valuable commodity? The answer is through the logical presentation of, and education to the applicable regulatory requirements and the benefits which can be derived therefrom. An important consideration in obtaining this necessary backing is the credibility of the individual or individuals advancing the QA auditing concept. The need for maintaining this credibility after program promotion and during program implementation is paramount. Loss of credibility will quickly translate into loss of management support and the start of a vicious downward spiral. Given top management support, the necessary funding and staffing requirements can be developed and approved.

The selection of personnel to staff QA auditor positions must be a rigorous, deliberate, well-thought-out process. The selection of capricious, argumentative, illogical, unprofessional or selfserving individuals can destroy an existing successful program, or delay for years the effective development of a burgeoning one. Auditors must foster understanding, maintain patience, be good public relations people, be capable of separating the chaff from the wheat, utilize others' time efficiently and judiciously, write factual, indisputable findings, and issue audit reports which stand by themselves, without need for oral expanatory statements by the auditors. Contrary to popular belief, auditors are not God - but, through the strength of knowledge and professional conduct, they can be a major influence on and benefit to the entire QA effort, management, the organization, and the end product.

I submit that it should be every auditor's fervent goal to some day be able to perform a thorough audit without finding a single significant finding. Although it has yet to be my personal experience to achieve such a revelation during an audit, I have heard the occasional isolated sounds of an auditor who has entered such pearly gates.

The notion that an auditor must have findings to justify his job is unacceptable to me. Auditors found to exhibit this personal philosophy should be exorcised from the program, auditees who fear such an approach should be counselled and assured to the contrary; and, programs which endorse such a backward concept should be scrapped, together with its misguided management.

PROGRAM'S BENEFITS

There should be little debate on the regulatory necessity for independent review of quality affecting activities through audits. The need to continually assure the protection of the health and safety of the public is a responsibility which cannot be overemphasized. It is vital in maintaining nuclear power as a viable, safe form of energy. However, I believe there are additional motivators for enacting a thorough, well-managed, professional QA auditing program.

One of these motivators is the realization that often the only direct interface a contractor or vendor has with the client, is during client's audits of their activities. This could be true whether it is a reactor vendor auditing one of its suppliers, an architect-engineer auditing a vendor, a utility auditing the architect-engineer or the Nuclear Regulatory Commission inspecting the utility. The potential impact of this "flag waving" opportunity is significant, since the image of the auditing organization could hinge almost entirely on the manner in which the audit team executes its responsibilities. The extent to which the auditors are professionals could affect the attitude of the audited company toward the project as a whole. Immeasurable harm or benefit can be derived from the experience - I have seen both. The positive reaction has never failed to impress and please me. The negative situations always make me want to shake the auditor by the shoulders to reemphasize the impact his actions have.

An effective audit can cause the audited organization to change its attitude in addition to its behavior. This is most important, for behavior only reflects what we do - attitude reflects how we feel about what we do - one's commitment, if you will. An effective audit can also place emphasis on what is important, and thereby reduce the amount of effort or attention being placed on extraneous or insignificant activities - activities which are not contributing toward the quality of the service or product.

A very important benefit of a comprehensive, well-planned, fully executed and documented program is its capability to satisfy the third party auditor who is auditing, inspecting or surveying your company.

If your own internal auditing efforts are weak or incomplete, expect the third party auditor to penetrate deeper into your company's activities. If your external audit efforts are inadequate, expect the third party auditor to consider accompanying you on an external audit, having you increase your audit frequency, and/or having other incremental impacts on your program. Conversely, a sound QA auditing effort will establish credibility with the third party auditors, resulting in more favorable reports on the company's management and activities; a very positive cause-and-effect relationship.

Management's satisfaction with a credible program may lead to the auditor's involvement in investigating "those sensitive" situations which occur from time to time in an organization. Whether this involvement takes the form of specific participation by the QA auditors or simply using their techniques, it further demonstrates that management perceives the QA audit process as an effective management tool which can be tailored to investigations, whether they be QA-oriented or not. The use of auditing techniques in typically non-QA areas is a vital theme. A proven auditing system can result in other organizational components emulating key features of the program, such as planning techniques, reporting formats, scheduling methods, briefing session philosophy, status and commitment tracking, etc. This form of flattery will increase the level of understanding and acceptance of the overall QA program, thereby strengthening the QA effort.

FLEXIBLE SYSTEM

Auditing is a flexible technique and as such can be easily applied to a variety of areas. Examples of this application are financial auditing, operational or management audits, administrative reviews to determine the facts surrounding particular situations, whether they be EEO discrimination claims, potential personnel disciplinary cases, environmental infraction allegations, safety investigations or other fact finding missions.

The Comptroller General of the United States notes in the forward to a pamphlet entitled "Auditors Agents for Good Government," that "Auditing is often considered... to be primarily concerned with the proper safeguarding of funds and property. Actually, this is only part of the concern of auditing."[1] The Comptroller General goes on to state that..."Auditing can be very useful to those...who recognize its potentials and use its reports. It can, for instance, alert them to potential problems so that they can make programs work effectively and correct inefficiencies and uneconomical practices before serious or even irreparable harm is done."[2]

During my association with the nuclear industry, I have seen audit concepts successfully function in such non-QA related situations as a review of conflict of interest charges, fraud allegations, assessment of cause and responsibility for a fish kill, compilation of facts regarding contract disputes, determination of need for specific management systems refinement, and so on.

Additionally, I have seen audits effectively performed on such non-routine QA subjects as a review of disagreements between surveillance agency personnel and the personnel of the manufacturing company being surveyed, a review of statements quoted in newspaper articles by contractor's personnel regarding QA, reviews of questionable material suitability charges, and others.

These experiences clearly indicate that audits are flexible, viable options which management can, should and frequently does rely on to determine the effectiveness of its operations.

SPECIFIC EXAMPLE

To further emphasize the usefullness of this management investigative tool, I would like to cite one example in particular. The case in point is the North Carolina Utilities Commission ordered audit of Carolina Power & Light Company (CP&L). This audit, or study, was an eight-month evaluation of CP&L's management systems, procedures, and performance. CP&L, a company whose hallmark is a progressive style, open communications and responsive management, issued a pamphlet entitled "What Kind of Job is CP&L Doing?" This document highlights the management performance study which they had undergone by an independent, outside consulting firm and provided its' customers, shareholders and employees with a concise, accurate summation of a 700 page report.

In the pamphlet, the purpose of the audit is stated as follows:

"The objectives of the study, as stated by the auditors, were to provide an impartial, professional assessment of CP&L's management and operating efficiency, to identify opportunities for improvement, and to provide recommendations for achieving improved, more cost-effective operations."[3]

Commenting on the report, CP&L's chairman notes in the preface to the pamphlet:

"The results of the audit show that CP&L is operating efficiently and that we are doing a good job for our customers. As in any large organization, CP&L has areas for improvement. Where the potential for improvement has been identified, we have initiated appropriate action."[4]

The pamphlet concludes by noting:

"The report pointed out that CP&L has continued to be responsive to major changes - in economic conditions, in regulatory climate, in technology -

---

[1] Auditors Agents For Good Government, Audit Standards Series Number 2, United States General Accounting Office, United States Government Printing Office, Washington, D.C., 1973, p. i.

[2] Loc. cit.

[3] What Kind of Job Is CP&L Doing?, Carolina Power & Light Company, Raleigh, 1977, p. 2.

[4] Ibid., p. 1.

and has been able to modify its organizational structure and operating functions, sometimes quite drastically, to adapt to these new conditions."[5]

In summarizing the findings, the outside auditors observed:

> "The interest in and resolve of the top management during the course of the study in moving directly and promptly on opportunities for improvement provides assurance to the consultants and the commission of CP&L's continued resolve to carry out these plans."[6]

The example of CP&L underscores the concept that audit programs can provide management with the opportunity to assure that vital activities are being performed satisfactorily, efficiently and in accordance with commitments, thereby maintaining the company's internal and external credibility.

MANAGEMENT PRINCIPLES

Another way of realizing the value of the audit function to goal-oriented management, is to select a few key phrases from criterion 18 of 10CFR50, Appendix B entitled "Audits". This approach results in phrases like "A comprehensive system....to verify compliance....and to determine effectiveness.... . Audit results shall be documented and reviewed by management.... ."[7] Aren't these the basic managerial principles of monitoring against goals, feedback, follow-up and management action?

In essence a QA program is a comprehensive management system, with audits representing certain basic elements of the overall management process. The concepts of planning, defining authority and responsibility, setting goals, integrating implementation, measurement, feedback and corrective action are all represented in an effective QA program; and can be reviewed through an auditing function.

Certainly audits are time consuming. But as the venerable Dr. J. M. Juran notes regarding audits in his book entitled "Quality Planning And Analysis," "All this takes time. This is the price one pays for the realism of facts over the drama of opinions."[8]

SUMMARY HIGHLIGHTS

In summary, to develop a useful, comprehensive QA Auditing Program, three fundamental ingredients are required:

- Credible QA personnel to advance program.
- Top management support.
- Selective hiring of QA audit staff.

Given the establishment of a well planned, professionally staffed QA Audit effort, program benefits can be realized in several areas such as:

- Owner's successful representation in outside firms.
- Improved attitude and behavior by others toward QA.
- Emphasis focused on salient points.
- Increased credibility with third party.
- Audit techniques used in non-QA areas.

In conclusion, I content that the limits to which QA auditing as a system are of benefit to management is quite simply a function of the credibility and resourcefulness of the QA people, and the commitment and perception of their management.

---

[5] Ibid., p. 9.

[6] Loc. cit.

[7] _Quality Assurance Criteria For Nuclear Power Plants and Fuel Reprocessing Plants_, Title 10, Code of Federal Regulations, Part 50, Appendix B, General Services Administration, Washington, D. C., 1977, p. 301.

[8] J. M. Juran and Frank M. Gryna, Jr., _Quality Planning and Analysis_, McGraw-Hill Book Company, New York, 1970, p. 563.

The Quality Assurance System Audit: Who Needs It?
Albin J. Rzeszotarski

FOREWORD

Successful business enterprises, especially those dealing with complex manufacturing processes, are finding that an effective quality assurance program (QA) is essential to the production of continuously acceptable products at competitive prices. A QA program includes all those activities which provide assurance that the operations are being performed in a manner which satisfies the quality requirements of the product. Whether these requirements are dictated by government regulations or are an edict of the marketplace, a means to measure the effectiveness of the quality assurance program is vital. A properly conducted quality system audit will accomplish this purpose.

A quality system audit is a periodic and systematic review and evaluation of the documented requirements of the QA program. An effective audit discloses any inadequacies in the interpretation of basic quality requirements and reveals failures in complying with procedural instructions. The cause of these discrepancies may also be determined. The audit is not only a means of verifying that the QA program is effective, it is also a very vital part of it, because in exposing program deficiencies, the program itself is improved.

TYPES OF AUDITS

Quality system audits fall into two categories-the program adequacy (or adequacy) audit and the compliance audit. Each uses a different audit base-the requirements against which a system is evaluated.

The adequacy audit answers the question, "Are the quality requirements we have to live by reflected in our QA program?" These requirements may be dictated by government regulatory agencies such as the Nuclear Regulatory Commission (NRC) or the Department of Defense (DOD). In the American nuclear industry, the overall quality requirements are specified by Appendix B of Part 50, Title 10, Code of Federal Regulations. Technical societies may also promulgate standards which are almost universally recognized and often levied as a requirement. One of the most common of these is the Boiler and Pressure Vessel Code established by the American Society of Mechanical Engineers (ASME) whose requirements on pressure boundaries, welder qualification and QA programs for suppliers working to the code are well known. The American National Standards Institute (ANSI), under the auspices of the ASME, has also developed standards for numerous activities. Customer or corporate requirements (or both) are incorporated into the QA Program. Finally, a comprehensive set of implementing instructions provides the details of accomplishing the various activities to make the program work.

The adequacy audit, therefore, reviews all of the quality requirements documents to determine whether they have been correctly interpreted by succeeding tiers of implementing instuctions. Therefore, these implementing documents must be continuously reviewed by the using organizations to assure that only the minimum necessary control is exercised. Requirements change, instructions become obsolete or inappropriate; procedures mature through changes resulting from the experiences of numerous organizations and individuals performing similar tasks.

The second category of audit is the compliance audit. The audit base is usually the lower-tier procedures and instructions. It is this category of audit which answers the question, "How well are personnel complying with the instructions they are supposed to follow?" This question is answered by a comparison of the documented results of quality activities, such as inspection reports, purchase orders, discrepancy reports, material acceptance reports, etc., with the procedures and instructions governing these activities. It is this category

---

7098 Bret Harte Drive, San Jose, California 95120

of audit which will be more fully described in this discussion.

How does the system audit accomplish its objective of improving the overall QA program? Perhaps the simplest explanation can be made through a detailed description of a generic audit- who the auditors are, how they determine what to look for, how they find what they are seeking and what they do about the findings.

AUDITOR QUALIFICATIONS

What constitutes an effective auditor? Jim Rusk of the General Atomic Company in San Diego, California, has listed the desirable attributes of auditors. These are personal integrity, responsibility, consistency, cooperativeness, constructiveness, adaptability, curiosity, persistency and common sense. Most of those traits are, of course, fairly obvious. Adaptability will allow the auditor to cope with systems which are new to him so that he may provide a more thorough and competent evaluation. The slightly curious individual may probe into some areas more suggested by "hunches" than by logic. The persistent auditor is not put off by the individual who is always too busy to be thoroughly interviewed. And the auditor must possess common sense or the audit may become completely impractical.

The members of the review team must possess the technical skills required for competent assessment of the areas they are investigating. They must have basic experience in the technique of auditing so that they can perform an evaluation which is sufficiently probing to disclose deficiencies which may not be obvious, but yet are significant shortcomings in the quality system. They must also be independent of the operation being evaluated lest the objectivity of the audit be jeopardized.

The following qualifications have been considered as requirements for lead auditors; Education, industrial and audit-related experience; professional accomplishments; training in auditing and passing of a formal examination; evidence of participation in successful audits as well as evidence of mature judgement and the effective communication of audit results.

Specifically, does the individual possess a degree from an institution in engineering or a related science? The level of diploma (associate, bachelor, master, etc.) should also be considered. Is the experience in the appropriate field, is it in our discipline (quality assurance) and, finally, is it in the auditing function?

How has the candidate displayed professional accomplishment? One of the most obvious means is by accreditation in the American Society for Quality Control as either a Certified Quality or Certified Reliability Engineer. Accreditation is granted after the successful passing of tests which examine the candidate's proficiency in all areas of quality or reliability. Additional evidence of professionalism is provided by the State of California which has recently inaugurated the licensing of qualified individuals as professional engineers in quality engineering. Formal training in auditing may be obtained, most readily, from company, society, or institutionally sponsored courses.

Finally, lead auditor qualifications include an examination of recent audits in which the candidate has participated. Does the individual exhibit the maturity of judgment necessary to obtain resolution of findings which may be highly critical of procedures and personnel? And lastly, is the lead auditor sufficiently effective in expressing himself to conduct all phases of the audit adequately?

Whatever criteria for the selection of lead auditors are used, the responsible manager must thoroughly evaluate the individual's capabilities and compare them to the task to be accomplished. The auditor must be examined as an individual and qualified accordingly. No certification is permanent, so that some period should be established for the re-review of qualified auditing personnel.

PLANNING THE AUDIT

We have determined who the auditors are. How does the auditor determine what to look for? Obviously by careful planning-the key to an effective audit. Planning must include details of participation by the audited party as well as the reduction of the areas to be reviewed to a detailed check list. Planning must include provisions for the proper level of management at the pre-audit conference, the required access to records and personnel to be contacted during the evaluation period, and the proper level of management at the post-audit conference. The proper level of management is that management which is responsible for the activities being audited (and thereby responsible for acknowledgement of the audit findings), for committing action to resolve those findings and for initiating corrective action to prevent recurrence of the discrepancies discovered.

The date(s) of the audit must be coordinated with the organization to be audited so that the required resources will be available to the team. Internal organizations may be audited annually while contracts with outside vendors should be evaluated for the contribution of their products to safety, profitability, schedule, etc. and the audit schedules established on that basis. The auditing organization should be evaluated by another group to maintain independence of the auditor-auditee relationship.

The audit notification should include a statement on the scope of the audit together with the audit team members, and identifying the team leader, the date of the pre-audit meeting, a daily audit schedule including those documents or areas to be reviewed and the date of the post-audit conference. A Quality Assurance Audit Plan is often prepared coincident with this notification to the audited organization and would include more complete information such as organization (s) and activities to be audited, applicable documents forming the basis for the audit, pre-audit information required from the organization to be audited and procedures/check lists to be used in the review.

The check list is compiled well before the audit and represents a detailed description of the items to be covered in the review. Numerous philosophies exist on exactly how each check list should be made out and it is obviously difficult to formulate the ideal pattern. Whatever the format, each appropriate portion of the instruction should be dissected from the whole as an audit item. For example, the instruction may state that all personnel performing a particular function have been adequately trained. This might be the "Audit Item", usually the initial consideration of the check list.

The next consideration on the check list might be entitled, "Audit Point". Here would be listed those actions (which can be verified by documentation) which must be considered if the audit item is completely implemented. In the example above, one audit point might be, "Is there documented evidence that predetermined criteria were used to determine the adequacy of training?" Another audit point might be, "Is there documented evidence that the education/ training/experience of the incumbent personnel has been evaluated against these criteria?" It is evident from the questions that a "yes" answer signifies compliance while a "no" signifies noncompliance. The questions should be structured this way to enhance the uniformity and clarity of the audit. The Audit Items and the Audit Points are prepared well in advance of the evaluation itself.

It is clearly evident that the auditors must be thoroughly familiar with the QA program and the documents being audited. Moreover, the auditors must be abreast of continuing changes in the program so that the specific items they are reviewing are current. They must recognize that there may be several means to implement a requirement and that the approach the audited party has chosen may be a valid one, albeit different from their experience.

In setting up this initial phase of the audit, the individuals (or the organizations) who can most appropriately respond to the audit items should be identified. This is especially important in those instances where an audit is comprehensive and involves diverse organizations and locations. Each team member must be utilized most efficiently to perform the review within a reasonable time with a minimum disruption of the activities of the audited organization.

In many organizations, because of the nature and size of the audited function, an audit by one individual may be adequate. In other groups, the diversity and/or complexity of operations dictate the use of the team as described earlier. Also, areas involving highly technical disciplines may require that the team be augmented by various specialists intimately versed in the area under investigation. Nondestructive examinations are in this category. The team approach however, should be utilized or at least seriously considered for all audits. Well-qualified members of a team can usually provide a more comprehensive review of the requirements than one individual, often with more penetrating and probing audit items. This is, of course, in addition to sharing the audit load.

For the initial audit of a facility, the planning portion of the audit, including the preparation of the check list through the audit points, normally may involve considerable time which should be budgeted for that purpose. Later audits, presuming a minimum change in requirements, can be performed more expeditiously. A standard check list may subsequently be developed from several repetitive reviews which makes planning simpler. The initial audit preparation almost always consume more time than the subsequent investigation.

Having determined who the auditors are and how they determine what to look for, how do they find what they are seeking? This, of course is the most interesting portion of the exercise.

THE AUDIT

This phase of the audit begins with the pre-audit conference. The audit team is introduced to the management of the organization being audited. The audit plan is discussed. The scope of the audit and the individuals to be contacted are confirmed. The purpose of the audit may need to be clarified. The manager to be audited is advised of the specific requirements to be covered in the audit. He should be encouraged to describe his organization and their activities. All arrangements for the cooperation between the auditors and the audited organization are finalized.

The compliance audit determines whether the control system used (the series of instructions) is being properly implemented. By definition, an audit is a sampling operation. Normally, because of the large number of documents generated in an activity, only a sample of these documents can be practically reviewed. The conclusions from the audit (how accurately does the sample represent the population as a whole) are as valid as the sample is representative. At this stage of the evaluation, bias can be interjected by the auditor in his selection of the audit sample.

A system of determining the audit sample must be followed which removes as much judgement as possible from the selection. A major danger in using a judgemental system is an unfamiliarity with all of the bases for making the selection. A supposed random selection of files from a drawer, for example, will almost always omit the first and last files; files may be ignored because of appearance or size; or the spacing of the files may influence selection. Additionally, all of the documents which should be sampled may not be in that file. Our judgement has not allowed us to objectively determine the proper selection of the sample. The results from auditing these samples cannot be utilized to describe the document population by objective methods; a statistically valid sampling method should be considered.

The binomial distribution may be used to determine the sample size required to find at least one violation at a given probability if the frequency of violations in all of the documentation is stated. Specifically, if 50% of the documentation is noncompliant, which might occur if a requirement is ignored and the requirement is met in the normal course of performing the operation, the probability of finding at least one defect in a sample of seven is approximately 95%. The size of the total documentation will have a relatively small effect on the probability figure even though the binomial is normally used for very large populations.

The total document population must be determined so that every item which is appropriate to the audit is included and has an equal opportunity to be selected. The estimated frequency of occurrency of violations in the population, together with the desired probability of finding at least one defect are established and the sample size determined.

Selection of the specific items to be sampled may be accomplished by using a table of random numbers. Each item in the population is assigned a sequential number, the starting point in the table is selected at random and the sample numbers determined by following a pattern from the starting point. The documents to be sampled are the third consideration of the check list and might be entitled, "Items Audited" and the sample selection method noted therein.

The analysis of the documents sampled, as well as the interviews conducted during the audit, may seem almost anticlimatic after the effort spent on the prior activity. Any deficiency noted from the requirements is verified with the individual designated liaison for the audit, or with responsible management. If interviews are required, the respondent should be asked to verify his statement by signature. If his responses are discovered by observation, the action observed should be witnessed by, and verified by, responsible management. It is extremely simple for the auditor to misunderstand and misinterpret the evidence he is reviewing with findings which may subsequently prove to be invalid; credibility of the audit is thereby put in jeopardy. Verification of findings is paramount.

The rapport established between the auditor and the interviewed party is extremely important in determining the outcome of the audit. Some cynics say that the sole output of the quality organization is criticism and that this is especially true of the auditor. This attitude will put the individual being audited on the defensive. He may feel his technical competence is threatened, or he resents interference into his realm, or he may have personal problems which cause his antagonism. The auditor can do little in many of these situations but be open in his approach, not display an attitude of arrogance, especially when discovering discrepancies, and try to portray an honest and cooperative image.

Some auditors may favor more extensive interviewing than others because, in addition to responding to the auditor's inquiries, the interview may disclose additional areas which the auditors seek to pursue. As with other evidence, the auditor should make the questions a part of his check list and verify (by signature) the responses received.

The results of the items audited are the fourth consideration of the audit and might be included as the "Observation/Finding" of the check list. The final entry might be marked "Comply? Yes/No". This provides simple reference to signify whether each item audited met the requirements. The noncompliant items are considered for the audit report as well as for corrective action.

RESOLUTION OF AUDIT FINDINGS

The final portion of the audit concerns what the auditors do about the findings. Significant deficiencies must be rectified, the need for purging of the existing deficiencies must be determined, and action must be taken to preclude recurrence of the discrepancy in the future. These three steps in correcting deficiencies are the distillate of the audit process.

Actions to accomplish the necessary corrections very with each finding. If, for example, the system requires that (prior to issuance) management review and approve drawings prepared by engineers, the audit should disclose how efficiently this requirement is followed. The audit may also reveal which engineers (and managers) are violating the instructions.

The cause of the violation may often be determined from the nature of the findings. If the use of unapproved drawings is general among several engineers, an unfamiliarity with the requirements (or refusal to acknowledge the importance of following them) should be suspected; an appropriate training program may correct this deficiency.

Repeated violations traced to one individual may indicate his disagreement with the requirements as stated (and therefore his refusal to abide by them) and may call for additional discussion to clarify the situation. In many cases, therefore, the "why" of the violation may be determined and corrected.

The formal document requiring corrective action resulting from the audit should include, first, a description of each significant finding, signed and dated by the Lead Auditor and secondly, an acknowledgement of the accuracy of the finding by the audited manager.

The Lead Auditor may then describe in a general fashion, the recommended action which will rectify the finding. Several arguments are offered against the audit team's suggesting the recommended action, the strongest of which concerns the prerogative of the affected manager to specify how he will manage his organization. A general description of how to meet compliance, however, may be helpful.

The affected manager's commitment to correct the deficiency mentioned in the finding must be stated in the corrective action document. The committed action statement must respond to the findings and not to the recommended action of the auditors. The manager must sign and date this statement, specifying a completion date for the action. The Lead Auditor accepts the committed action and completion date as satisfying the findings by his signature and date. Any differences between the audit team and the audited organization should, of course, be resolved by the appropriate level of management.

After all of the corrective actions have been finalized, a post-audit conference is held by the audit team with the management of the audited activities. The findings and subsequent corrective actions are discussed, as well as any concerns the audit team may have which were not worthy of formal corrective action.

Following the post-audit converence, the audit report is distributed to the audit team and their management and the management of the audited activity who will be able to perform the committed action to rectify the findings. A typical audit report may contain the identification of the audit team, a description of the audit scope, identification of persons contacted during the pre-audit, audit, and post-audit activities, a summary of the findings and committed action, an audit digest which provides details of scheduling, planning and conduct of the audit, and a copy of each corrective action document prepared as a result of the audit.

As the committed action in each corrective action document is accomplished, documented evidence of same is provided to the Lead Auditor. If the evidence satisfactorily meets the committed action requirement, the document is closed.

If verification of committed action requires a reaudit of the facility, this is accomplished in a later visit, often during the next audit.

CONCLUSION

The title of this paper asks, "Who needs a quality system audit? What does it do for me?" An effective adequacy audit will determine whether the implementing instructions adequately meet documented quality requirements. Adequate interpretation includes only those instructions which are necessary and excludes superfluous controls which may consume valuable resources of the organization. The compliance audit verifies that the procedures the organization has developed as the standard method of accomplishing various functions are being followed, that whatever quality system controls the organization has imposed upon itself are functioning adequately.

The audit, however, is not a substitute for any of the systematic controls of the QA program. If it is intended to have the quality organization review various documentation such as procurement orders or design specifications, for example, this requirement should be codified as an established practice and its implementation enforced. Its enforcement cannot be lax because of the excuse that the audit will pick up any violations. It is true that an effective audit, even though it is a sampling process, should expose these system violations. It is also true that once the deficiency is discovered, all of the appropriate documentation may then have to be examined to determine the extent of the problem. In any case, the required review is finally performed after the fact, with perhaps more of an expenditure of resources than if it were accomplished in real time. And, of course, the consequences of omitting the original review may be considerable.

In both audits, the efficiency of the operation is improved through the use of realistic procedures, uniformly followed. This is the purpose of the system audit. Until we have a perfect system, we can benefit from its audit; experience has shown that the benefits in improved efficiency of operation are well worth the effort.

Reference: Rush, J.H. <u>Quality Assurance System Audit</u>, San Diego, California, General Atomic Company, August, 1977.

AUDITING

Quality Audits in Support of Small Business

by Walter Wilborn
Associate Professor
University of Manitoba

In Manitoba, an effort to upgrade quality control in small business firms receives an enthusiastic response

Inadequate planning and control of operations ranks as a major cause for small business failures. This critically weak management extends to quality control. Many owners/managers do not consider quality of design, conformance, and performance an issue that urgently requires their attention. In a prevailing environment of informality and personal relationship, one still assumes "good workmanship" sufficient for assuring quality, although often it remains undefined and, therefore, uncontrollable. Technological advances along with other socio-economic changes have changed needs and conditions for quality control in small enterprises drastically.

We cannot expect that small business will improve its quality control without aids from the community. In realization of the importance of a healthy and competitive small business section in a free enterprise economy, various kinds of management assistance programs already exist. The Small Business Administration (SBA) (1), as the best known supporting agency, for instance, has helped effectively. In the area of improving quality control, however, the available assistance is minimal, mainly because small business operators do not perceive the need for, and the benefit of, modern methods. For outsiders, free enterprise principles do not allow undue interference with management. Still, in this case, initiative had to come from elsewhere than the small business, at least in Manitoba.

Quality audits appeared to us as a feasible and useful external aid to small business management. Before designing a suitable audit service we explored these questions:

- What is a workable definition of "small business," and what is its scope?
- What are unique features of quality control in small business?
- What are the limitations of current quality audits in small business as a guide for management?

SBA defines "small business" as "firms which are profit seeking, have 500 and fewer employes and are not dominant in their field." 97 percent of all U.S. companies fall into this category (2).

In recent years growing importance of quality control in small enterprises arose from:

- technology related to products, processes and equipment becoming more complex and demanding to producer and consumer;
- customers requiring compliance with generic standards for quality control programs;
- other small companies already establishing effective quality control procedures and gaining competitive advantage;
- legislation for consumer protection, higher incidence of product liability claims, and reduced rates for insurance requiring evidence of an adequate quality control program.

A manager/owner finds little help from his traditional management advisor, the accountant, with regard to quality control. Quality auditors that represent government and customers' interests cannot act freely and independently. These quality audits are, for the small business operator, of limited value since:

- Main purpose of the audit is to establish evidence and to attest rather than to assist.
- They are not requested by the owner/manager in awareness of need for, and possible benefits from, improved quality control. The audits are more or less imposed upon them.
- They are conducted infrequently, in conjunction with a specific contract or legal act, and without due regard to small business conditions.

Caught in the Middle

On one hand, mounting pressures for improved quality control, and insufficient service on the other, caught small business in the middle. It became obvious that the main problem rests with the owners/managers in that they are reluctant to act, and to initiate assistance. In order to overcome this mainly psychological problem we proceeded very cautiously (3). In our strategy and audit service we remain cognizant of the following characteristics of typical owners/managers of small enterprises:

- They tend to operate by "common sense," tradition, trial and error and day to day problems. They are not professional managers.
- They are hard working, self-reliant and energetic individuals and do not accept advice from outsiders easily.
- They maintain close personal relationships with employes, business partners, and the surrounding local community.

With these traits in mind, we arrived at a strategy by which we approached owners/managers first on a community wide basis and later individually. A "community quality audit" precedes and prepares "small company audits."

Main purpose and objectives of this new kind of quality audit are:

- generating information on the status quo of quality control, scope for improvement, need for assistance and availability of resources;
- creating an awareness for quality control in the community and its small business section;
- establishing first contacts with small business operators and exploring the clientele for company quality audits; and
- conducting research and provide a report for a steering committee.

Community quality audits constitute "audits," and not just "surveys," because they aim at:

- establishing a general standard for a community wide quality program and quality control improvement program;
- appraising and improving the effectiveness on such programs;
- implementing the policies of the steering committee;
- ensuring that human, financial, and material resources are adequately

known, utilized, and developed; and
- initiating remedial action and follow-up audits at both community and company levels.

System Oriented

With regard to the classification of quality audits by Mills (4), community quality audits describe a location and system oriented audit rather than a product and process audit.

Who should initiate such community quality audits? This responsibility will fall on the steering committee having a strong representation from small business. While government officials will not qualify as independent auditors, they can contribute as committee members along with those of other organizations.

The baseline audit sets the stage for the entire audit program. Actually, anybody significantly motivated, informed, and able to attain the cooperation of others can pioneer such an audit in the local community. The author, being involved in the local quality movement for some years as educator, executive, and member in the ASQC Section of Manitoba helped originate the first community quality audit. The Manitoba government had conducted similar so-called productivity audits and joined in the project along with individuals from the university, ASQC, and other local organizations. A faculty member of the local university has particular advantages to act as "neutral" leader. In the U.S. such audits qualify for grants under Bill S-972 for small business development centers at universities.

Questionnaires

In our Manitoba audit we applied a fairly conventional approach. We used questionnaires and question checklists. Of 700 small manufacturing establishments, about 40 percent responded. We checked a great number through subsequent interviews. These we carefully prepared as we wanted not only to gather data, but at the same time, establish first personal contact with a hopeful future client for a company quality audit. With the exception of the initially mailed questionnaire the procedure compares with that of a normal company audit.

We tabulated the answers and statistically evaluated them for proper analysis. Most interesting and revealing were also answers to our open questions. It became quite clear that many of our owners/managers recognized the importance of quality control. They want better training opportunities and information service in order to overcome quality control problems. With regard to company quality audits, many respondents were prepared to submit their current practices to an appraisal as a prerequisite for improvement. These audits, however, were to be conducted by an independent, well qualified auditor, who is neither employed by another company nor the government. At the end of the community quality audit we knew our potential clientele for company quality audits quite well.

We forwarded a report to persons in local associations, government departments and educational institutions, along with a request to join a steering committe. All of them responded enthusiastically, somewhat to our surprise. This showed that, at least in our province, quality control is an issue very much on the minds of many people. This independent "citizen committee" elected the author as its chairman and provided a well appreciated and unique forum for discussing concerns for better quality control in the business secion of the local community. With focus on improving quality control, the committee developed its function and role as a policy making and managerial body. (See Figure 1.)

The report from the community quality audit provided the major basis for discussion and constructive recommendations. As a result of the quite demanding work of the committee a government position for a coordinator of quality assurance has been created, a university certificate course in quality assurance management arranged... with a surprisingly high enrollment, and, finally, quality audits in small companies have commenced.

The following reasons caused owners/managers to participate in quality audits:

- Through the preceding community wide project they simply wanted to continue in a project with which they identified themselves already; they continued in the effort.
- They learned from meetings and the report enough about quality control with regard to their immediate local environment in a most realistic, simple, and practical manner. Naturally, they wanted to know more details about the comparison of their current company practices with the prevailing status in the surrounding local community.
- They acknowledged and appreciated the steering committee and its individual members as competent and reliable trustees for conducting and monitoring company audits and independent quality auditors.
- They became acquainted with some of the quality auditors who actively participated in the community quality audit.
- They realized that benefits will easily exceed the costs partly because public grants for the audit were made available.
- They recognized that, because of the prevailing close personal ties in small business, bringing about more formal and restrictive quality control procedures would be difficult without the help from a competent quality auditor as an outsider.

In more detail, owners/mangers expect from this company quality audit:
- an interpretation of the report on the

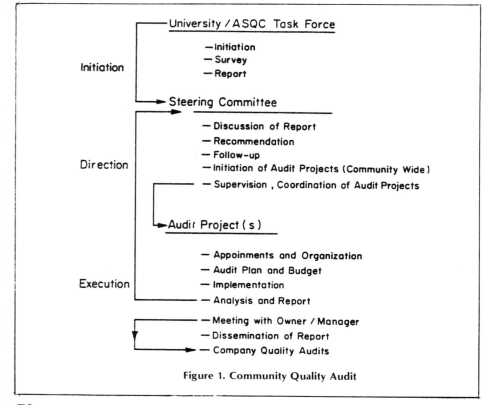

Figure 1. Community Quality Audit

community quality audit with regard to the individual company;

- an assessment of current quality control practices of the company and recommendations for simple, practical, and short-term improvements;
- assistance in setting up such an improved quality control program;
- documentation of this quality control program in an easily understandable language and form, that facilitates implementation and maintenance through self-audits by the owner/manager; and
- further convenient availability of the auditor whenever needed.

The owner/manager does not want an audit that completely by-passes him/her and results in a report. Obviously, a quality audit geared to assist a small business operator has different features when compared with those in large companies. The auditor of a small and independent enterprise must work very closely through all phases of the audit with the owner/manager in order to be accepted and to render useful service.

Consequently, such an auditor must have acquired additional qualifications such as knowledge and experience in small business management along with a sincere appreciation of small business values; ability to communicate well with small business operators, to listen, to understand, and to work with them rather than for them; and competence in adopting generic program standards, specific procedures and methods to needs and conditions in small business.

From the discussion of the community audit report and the owner/manager's perception of quality control in the company should emerge fairly precise expectations and goals for the audit. Such a careful preliminary orientation of both parties ensures best the owner/manager's continued interest and active participation in the audit itself. The audit is built around the owner/manager entirely. A final audit plan shows various activities and their purposes and also a time schedule that allows full participation of client and auditor. An audit plan, formulated in writing, constitutes a contract that can assist also in raising financial support for the project.

Ten Steps

In order to guide further research and development in small company quality audits, we designed an audit plan as a kind of tentative standard. If the owner/manager requests a quality audit without the involvement in a preceding community audit, this company audit plan can still be applied. This standard audit for a small enterprise achieves the actual establishment of a practical and suitable quality program through the following steps:

1. Determine existing quality control/inspection practices.
2. Formulate these in writing in a simple language and augment with charts, etc.
3. Discuss, clarify, and possibly revise these current practices with owner/managers and other individuals concerned.
4. Document approved practices as standard procedures in a quality manual.
5. Determine voids by comparing current practices with needs and possibly suitable generic standards for quality control programs. Have owner/manager decide whether additional control function and procedures are required.
6. Prepare simple forms for reporting and data compilation and document these in the manual.
7. Implement procedures and reporting schemes; possibly revise when necessary.
8. Prepare checklists and simple plans for self-audits to be conducted by the owner/manager or an employe and indicate these in the manual.
9. Set date for next self-audit.
10. Clarify future services by the auditor and enter auditor's endorsement of the quality programs in the manual.

These ten steps would have to be further extended in order to suit individual circumstances and personalities.

The standard audit plan once jointly executed provides a most valuable learning experience for both partners. Owners/managers become adequately acquainted with modern quality assurance planning and sound management practices. The focus on learning in an extremely practical and mutually beneficial manner strikes the main difference between this small business quality audit and others. Usefulness derives from the auditing exercise, and not from a final report and recommendations.

Moreover, the tangible outcome, the quality manual, represents and documents an operational and actually implemented and hopefully proven quality program. The owner/manager has prepared this manual and it constitutes most likely a real management breakthrough. Without question, this program has every chance to be properly maintained through self-audits, and improved when needs command this.

References

1. *Small Business Act*, Public Law, 85-536, U.S. Government Printing Office, Washington, D.C.
2. Peterson, R., *Small Business, Building a Balanced Economy*, Press Procepic, 1977.
3. Vasallo, J., Psychology of Auditing, *Quality Progress*, September, 1977, p. 37.
4. Mills, Ch. A., In Plant Quality Audit, *Quality Progress*, October, 1976, p. 22-25.

About the Author

Walter Wilborn is associate professor at the University of Manitoba, on the faculty of administrative studies. He came to the university in 1968, after teaching various business subjects in other schools in Canada. Quality control is one of the subjects he currently teaches. He has done research on quality assurance programs, generic standards, and audits.

Wilborn is chairman of the Steering Committee on Quality Assurance Development for the province of Manitoba, a member of the Task Force on Quality Audit for the Canadian Standards Association, and a member of the Technical Committee on Quality Auditing of ASQC.

He attended the universities of Innsbruck (Austria), Hamburg (Germany), and Manitoba, and holds a PhD in economics.

Patrol Inspection: Guide to Productivity & Profit

James E. Cooper
Quality Assurance Engineer
Westinghouse Electric Corporation
Hunt Valley, Maryland

In many companies and corporations, the inspector is looked upon as the individual who causes headaches to many people at the end of the manufacturing line. If a scheduled ship date is missed because the inspector is trying to do an adequate inspection job, the production and schedule people fuss and fume and try to appease the customer. If material has to be scrapped or reworked because of discovered deficiencies which cannot be easily corrected at this late stage of the manufacturing cycle, cost management people scream that all inspection does is run up cost by causing material to be rejected. If the customer complains about defects which he has received, the inspector will get it from all sides because the defects were not caught prior to shipment. The inspector's job is a thankless one; he is expected to be the policeman in the midst of heavy traffic. With this kind of conditions facing him, we find that the inspector is constantly skating on thin ice.

What, then, can be done to assist the inspector who finds himself in this situation? One method, which has been implemented in our plant, is the setting up of an effective patrol inspection system. What is patrol inspection? Very simply put, patrol inspection is the process of getting the inspector away from the end-of-the-line and getting him to do more in-process inspection during the manufacturing cycle. By use of this technique, many of the accept/reject decisions at the end of the line can be eliminated. The backup that normally occurs in inspection also can be eliminated completely or reduced to a manageable level.

A practical example of patrol inspection versus end-of-the-line inspection can be cited in the building of a house. Those of you who have built a house, as I have recently done, can readily understand the benefits and the practical value of patrol inspection. You can check on the work of the builder from the digging of the foundation, through the installation of walls, plumbing, electricity, and cabinets, to the final completion of the building. You can have defects corrected immediately and less expensively by monitoring the work in process. These benefits are not available to the individual who comes along and buys the house after builder completion. His end-of-the-line decision is whether to accept or reject the house after completion in the same manner as the end-of-the-line inspector does with his product. Much of the complex and/or preliminary work is hidden from view of the inspector . . . and defects are not detected as readily. Also, the repair of discovered defects is more costly in an operation with only end-of-the-line inspection.

What do we, as members of the Quality Department, have to do to establish an effective patrol inspection in our respective operations? First, we have to ensure that *all* personnel who are involved in the manufacturing cycle are adequately and completely prepared for the institution of the patrol inspection. The inspector, obviously, needs to be properly trained since he will be the one responsible for performing the area patrol inspection. He needs to know what is required and how it is to be accomplished. However, in our plant we have found that proper training of the inspector is not enough. Material people have to be advised in the proper methods of storage and handling of materials to avoid potential problems and rejections which may occur late in the manufacturing cycle. Metal scratches, component leads shorted together and printed circuit boards piled on top of one another with no protective material between them, are examples of poor material handling. Such poor handling could have been prevented by more careful handling by the material people, or by looking at the material before issuing it to manufacturing. These conditions, resulting from poor handling, could cause material to be scrapped or reworked at a much higher cost.

Manufacturing personnel also have a responsibility in the patrol inspection operation. If a hand tool is not functioning properly, they should not put the tool back on the shelf and use another one. The defective tool should be placed in a conspicuous location and the operator should notify his supervisor and the inspector to have the tool rejected and repaired. If an operator notices that a machine is producing parts which obviously are not according to specification requirements, he should stop the machine immediately and reset it rather than wait until several hours have elapsed before taking corrective action. Training manufacturing personnel in the function of patrol inspection and the role that they have in it, will lead ultimately to the operator inspecting his work more frequently prior to inspection submittal.

Quality engineering personnel also have to be made aware of the tool that they have available to them, and how to interpret data supplied by the patrol inspector. Through this training, all personnel will learn what is required of them.

The next step should be the establishment of written guidelines or checklists to aid the inspector. Areas that are included on these lists should relate to the operations to be performed in the area. If there is a peculiar process operation to be performed, or if an assembly requires special instructions to be completed, such instructions should be included in the inspector's information and should be monitored by him. These guidelines, depending on the nature of the product involved, may be of a very specific nature, such as witnessing a torquing operation to 10 inch-pounds; or may be general in scope, such as ensuring that all hardware be installed properly. Whatever the definition, formal documentation of the patrol inspection guidelines is a must.

Now that the inspector has been trained to do the job and has been given the proper guidelines for his operation, a frequency of inspection is required. This depends again on the nature of the product involved. Some products and processes need to be checked hourly and others need not be checked more than once a day. Whatever operation involved, however, should be checked at some specific interval. You would not want a machine that turns out 60 parts an hour to run unchecked for eight hours; then find that something had slipped, and all parts made in the last four hours were oversized by twenty thousandths. By the same token, an operation on a large project such as a generator rotor may not need to be checked more than once a day.

What we have done is to include in the checklist an audit frequency which provides the inspector a ready reference as to what is to be checked and how often.

With all of the preparation behind us, we are now ready to reap the benefits of patrol inspection. The first benefit is the fact that corrective action can be implemented as soon as the deficiency is detected. The patrol inspector should be describing any problem to the responsible foreman. At the same time, a written report which requires corrective action within a specified time period is handed to the foreman. This written reminder becomes a part of the quality history, and can be used to determine trends attributable to operator, machine, or other factors. Problems which are repetitive in nature can be detected by the patrol inspector and corrective action can be implemented before the problem has gone too far.

Secondly, with the implementation of corrective action early in the cycle, scrap costs are going to decrease. There will not be as much scrapped material at the end of the line because the costly defects which were previously causing scrap will be corrected sooner in the manufacturing operation cycle. Also, the costly reworks due to end-of-line detection of defects will be reduced.

As the inspectors go about their jobs, they will get to know the production people and the type of work that they produce. This will help to establish confidence levels in these people. The inspector will learn that the work of certain people requires a great deal of in-process checking, and the work of others, not as much. The inspectors will become more effective and will be able to use their time more effectively by giving more attention to problem areas. The production people, in turn, will start to feel that the inspector is there to assist them in producing a better quality product. They will start to show the inspector those items that are bad and those items that are marginal. This then reinforces the operator confidence levels that have been previously established in the minds of the inspectors. Also, some production people will begin to get involved in patrol inspection. As stated previously, some operators will start to inspect their own work while still in the manufacturing cycle.

In addition, the backup which occurs at many final inspection stations is lessened because the inspector does not have to do as detailed a review of the submitted material; he has been following it through the manufacturing process.

Finally, an impact will be apparent in Quality Costs. Until the system is fully implemented on all material in manufacturing, the appraisal costs of inspection will increase because of a dual inspection effort. However, this increase will soon be reduced to the normal level of appraisal cost that has previously existed. The area of warranty costs will decrease because fewer defects will be delivered to the customer. This results in happier customers.

Consideration must be given as to how much authority the inspector should have in shutting down a line or a process where there is evidence that shutdown is warranted? Some companies give the patrol inspector complete authority to stop the line. Others say that the inspector's responsibility ends when the line supervisor is notified. Any problems after supervisor notification then become the responsibility of that particular supervisor. In our company, the supervisor and the responsible Q.A. Engineer are notified of the problem by the inspector. Together they review the problem and determine what action is to be taken. This is something that should be determined for each individual operation when patrol inspection is instituted.

What all of this boils down to is that through the implementation of Patrol Inspection, we are getting more productivity, first from our inspectors, and then from our production people. Increased productivity, in turn, produces increased profit for our employers.

We realize that the objective of all quality assurance personnel is to attempt "to thoroughly analyze all situations, anticipate all problems prior to their occurrence, have answers for these problems, and move swiftly to solve these problems when we are called upon . . . however, when you are knee deep in alligators, it is difficult to remind yourself that your initial objective was to drain the swamp."

Patrol inspection may not be the long awaited utopia, but it does help us begin to drain the swamp.

CHAPTER 5

QUALITY COSTS

Reprinted from INDUSTRIAL ENGINEERING, February 1978

Using quality cost analysis for management improvement

A method is devised to show how changes in operations and/or personnel, based on cost of quality, will benefit management.

Lee Blank, P.E., GTE Data Services, Tampa, FL
Jorge Solorzano, El Paso, TX

Most of the decisions concerning management personnel, organization structure, and promotion recommendations do not take product quality into account. Once the people have been selected to fill the positions in management and quality control, it is generally assumed that the quality assurance program will operate efficiently and economically. In fact, even if the company has a long-standing high service record in product quality, the cost to the company and customer for this level of quality is not of great concern to management or users.

Quality cost study

A quality cost system has been defined as the business of assigning a monetary value to things that go wrong[1]. That is, a quality cost analysis is done to determine how much it costs to maintain a certain level of quality in the manufactured items or services of the company. The quality costs are collected and categorized so that the specific purposes for the costs can be determined and analyzed. Juran and Gryna[2] and the American Society for Quality Control[3] have presented data collection approaches.

The categories used for quality cost analysis are defined in terms of quality assurance functions, rather than with specific regard to departments in the organization. Therefore, there will undoubtedly be costs which are classifiable as quality costs that originate in departments other than quality. The two main quality cost categories are discretionary costs and consequential costs.

Discretionary costs—These are all sums of money which must be spent to *assure* that a desired level of quality is present in the service or finished product. There are two types of quality costs which may be incurred at the discretion of the management. They are: 1. Prevention costs: Management must decide how much to spend to prevent quality problems which degrade product quality below a minimum acceptable level. 2. Appraisal costs: These costs are incurred to determine the actual quality level of the finished product prior to its placement on the market.

Consequential costs—These are sums of money which must be spent to *correct* the service or product, because the quality level is too low. These costs often occur because management did not avoid them by incurring a discretionary cost. The categories are: 1. Internal failure costs: These are costs which must be spent to improve the quality level of the product *prior to delivery* to the customer. 2. External failure costs: These costs must be incurred to correct all the product found to be below some minimum quality level *after delivery* to a user.

Figure 1 is a simplified sketch of a production and inspection system. The main sources of each category of quality costs are shown. Only prevention costs originate from all components of the production system, since it is everyone's responsibility to prevent quality problems, but only some people's responsibility to discover quality faults.

Of the costs defined above, prevention is one of the most important because it can be used to reduce total quality cost. However, due to its less specific definition, prevention expenditures are harder to quantify than other quality costs. Therefore, the actual data collection may be done in the following order: 1. Appraisal costs. 2. Internal failure costs. 3. External failure costs. 4. Prevention costs. This order is suggested because with it the engineer can use the more easily collected appraisal (inspection) costs as a learning mechanism to classify data, how to read company accounting records, and who to go to in the company for quality cost information.

Assessing quality costs

A more detailed discussion of what types of costs fit into each of these four categories follows. You may, in practice, want to refer to References [2] and [3] for further definition and collection techniques of these costs.

Appraisal costs—These are usually the largest quality expenditure for a

Figure 1. Sources of quality costs by category.

company because management commonly feels (consciously or unconsciously) that more inspection and testing will undoubtedly improve quality. Appraisal costs do absolutely nothing to prevent or correct defective product. They are spent only to discover that it does exist. Characteristically, a quality cost study will indicate that as much as 50%, and, if you have a quality-minded management, as little as 10% of quality costs are appraisal.[2,4]

Data collection here is fairly simple. If you determine the following you will have most of the quality appraisal costs:
- Pay for all full-time and spot-check inspectors, supervisors, testers, and clerical workers in incoming material and in-process inspection.
- Costs to own or lease and maintain all materials and equipment used for inspection and testing of product.
- Cost of processing and evaluating all appraisal data.
- Cost of all quality audit efforts on in-process or finished goods.
- Costs to perform field tests.

Internal failure costs—The expense to correct defective product prior to its release is usually about one-half of the total failure cost (internal plus external). You can usually anticipate that a discovered and corrected internal failure will be less expensive than an external failure, because the customer is not involved at all with the internal failure. Since the cost of product liability has risen so dramatically in the last few years, it is more important than ever to prevent or find a major portion of defectives prior to release from company control.

Since appraisal costs are incurred only by the discovery of production-generated failures, internal failure costs are also incurred in the production system to remove the mistake in the most economic way. If the following costs are determined, most of the internal failure costs will be accounted for:
- Cost of all items scrapped because they miss quality specifications.
- Total cost of reworking (completing some missed production operations) and repair (correction to salvage the product) for defects found prior to or during appraisal.
- Expenditures incurred by failure analyses on company and vendor materials and processes.
- Costs to re-appraise product which has been reworked or repaired.
- Loss in revenue due to reduced price for product of poor quality.

External failure costs—This cost may be large for two rather unrelated reasons. First, the company must bear the expense of correcting, to the customer's satisfaction and the company's ability, the failures discovered after release of the product from company control. Second, lost revenues because customers will not purchase the product due to their, or another customer's, past bad experiences. The first cost can be estimated from records by thorough data collecting, but the second falls in the area of loss-of-goodwill and is quite difficult to quantify. The total of internal and external failure costs is usually 50% to 90% of total quality costs.[2]

External failure costs are more difficult to obtain than internal failure costs. However, the following components include most expenses:
- Costs of processing, repairing, and replacing product returned for quality reasons.
- Costs incurred for customer contact and service to handle quality complaints.
- Costs of replacement during warranty period.
- Total cost of testing, legal services, settlement charges, etc., for product liability problems.

> **Lee Blank** is a Management Scientist with GTE Data Services. His primary areas of interest are quality and inventory system management and engineering economics. He obtained degrees from St. Marys' and Oklahoma State Universities and has served as associate professor of IE at the University of Texas at El Paso. He has conducted seminars in quality management in Mexico and the US. Dr. Blank is a Senior Member of AIIE, a member of ASQC, and a registered professional engineer.
>
> **Jorge Solorzano** has worked in the areas of production engineering and quality control for US electronics firms with plants located in Mexico. He received a BS in Electrical Engineering and an MS in Engineering Management from the University of Texas at El Paso.

Prevention costs—The actual dollar values spent in prevention of quality problems are quite difficult to obtain. Estimates must be made, especially when quality costs are being collected for the first few times. Prevention costs occur in several departments because many different types of personnel assist in the avoidance of future defective material. The first quality cost study will probably be a relatively poor estimate of actual prevention costs. However, do not be surprised if costs in the category seem very low when compared to other costs, especially appraisal. Historically, it has been found that prevention is only 0.5 to 5% of total quality system expenditures[2,4]. The percentage is low, not because prevention costs are so hard to collect that much of them are missed, but because such a large proportion of the quality dollar is spent trying to find and fix failures,

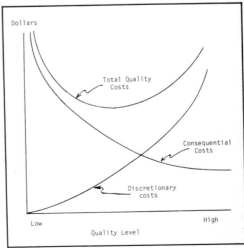

Figure 2. Effect of discretionary and consequential spending on total quality cost.

instead of preventing them. The majority of prevention costs will be included if the following are determined:
- Costs and salaries to plan better quality by engineering, process control, training, etc.
- Costs of designing and developing equipment, processes, and systems to measure and control quality.
- Expenses incurred in all types of quality training.
- Costs to improve quality by special studies, vendor relations, data analysis, etc.
- Costs incurred by departments other than quality assurance to improve quality.

Increased prevention costs?

Prevention costs are usually the smallest category and the most neglected, because management finds it difficult to evaluate the profit results of planning, trouble shooting, designing, communicating, etc. It is much easier to *count heads*, as it were, and thereby increase total quality costs in the categories of appraisal and failures. The expenses incurred in the name of prevention offer a large amount of quality-leverage, because prevention is so much more powerful and long lasting for the quality image of the product. For example, if prevention costs are incurred to find and correct a quality-morale problem among assembly personnel, the quality of all future product is improved. On the other hand, if appraisal costs which are

Case study

Since many industrial engineers and, in fact, most managers are new on the quality cost scene, a basic cost collection and analysis case study is presented. Assume that you are a new plant manager in an electronics product manufacturing plant located outside the US, but you report organizationally to the main US office. You know that there has been quality deterioration at the plant, that there have been virtually no quality control personnel changes, and that previous management had not been able to correct the situation. What would you do to improve quality? This case study is an example of what should happen if you had a quality cost study performed.

Except for plant management, most of the employees are nationals. The organization chart for quality control is given in Figure 3. Of the two supervisors of product acceptance one is in charge of process control inspection and the other supervises all final product inspection and testing. The supervisor of salvage operations manages all personnel who determine the cause of detected internal failures. He is also responsible for rework of these products, because these operations have been separated from the main production line due to high scrap rates. The economics of this arrangement are not studied here, but the arrangement makes quality cost data collection simpler.

Since this is a foreign operation, once product is shipped stateside it would not be returned to the plant in the event of an external failure. Repair is done at a US location for all manufacturing locations, and charges are not maintained for each plant. Therefore, there are no external failure costs that can be used specifically in this analysis.

Data collection

To collect the pertinent quality cost data the following things were done[5].
- Objective of the analysis—The objective of the analysis was determined as follows: To observe trends in cost data (for each category and total) and to relate these

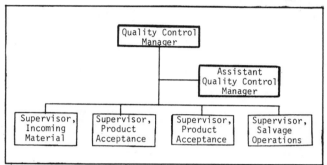

Figure 3. Organization chart for quality control.

Quality Cost Category	Costing Purpose	Percent of Time for Each Purpose				
		QC Manager	Assistant QC Manager	Supervisors, Product Acceptance	Supervisor, Incoming Material	Supervisor, Salvage Operations
Appraisal	Incoming	35%	10%		90%	
	In-Process/Product			40%		
	Audits		10	10		
	Data Analysis	5	10	10	10	
Internal Failure	Rework/Repair	5				60%
	Scrap					40
Prevention	Planning	30	40			
	Process QC			30		
	Training	10	10	10		
	Miscellaneous	15	20			
Total		100%	100%	100%	100%	100%

Table I. Time percentage breakdown for quality cost purposes.

Purpose	Function	Percent of time	Monthly cost	Monthly total
Incoming Material Testing and Inspection	Manager	35	$ 560	
	Asst. Manager	10	80	
	Supervisor	90	495	
	Technician	100	310	
	6 Inspectors	100	1,320	$ 2,765
In-Process Testing and Inspection	QC Supervisors	40	$ 440	
	3 Technicians	100	930	
	20 Inspectors	100	4,400	5,770
Set-up for Testing and Inspection	Eng. Manager	40	$ 160	
	3 Engineers	40	920	
	Technician	40	124	1,204
Quality Audits	Asst. Manager	10	$ 80	
	QC Supervisor	10	110	
	Technician	100	310	500
Equipment Calibration	2 Engineers	10	$ 175	
	5 Tech. (testing)	10	155	
	2 Tech. (production)	30	186	516
Inspection Data Analysis	Manager	5	$ 80	
	Asst. Manager	10	80	
	2 Supervisors	10	100	
	2 Analysts	100	500	770
Materials			$ 200	200
Total				$11,725

Table II. Summary of appraisal costs for October, 1975.

also discretionary, are incurred to correct the problem, internal failure costs, which are consequential, will usually increase. But, and this is important, morale will be further decreased because: 1. The defective product must be repaired. 2. The employee's and management's dissatisfaction with each other will increase. 3. The entire quality image of the company will suffer. This spiral effect can only be broken with time, effort, and discretionary expenditures to *prevent* further quality and morale reductions. The generally observed relation between discretionary, consequential, and total quality costs is shown graphically in Figure 2.

Management has a definite responsibility to determine how much the company will spend to improve quality. This decision should be made consciously using the advice of quality assurance personnel. The spending level should be reviewed periodically by responsible management. The final decision on future prevention spending should be made by management personnel who are organizationally above quality control, because they will be able to make the decision from a more informed position. Besides, this review of the scope of the quality commitment will make management more conscious of quality assurance's role in the company. The management should conclude that, even though prevention costs may be harder to justify and measure, these costs will do much to reduce the

---

results to management turnover and style. This objective was selected because turnover in management positions had been larger than expected.

- Time allocations for all quality personnel—The percentage of time spent by each person was categorized as appraisal, internal failure, or prevention. Table I presents this breakdown for all supervisory and management personnel. The breakdown is very important since it is used to proportion quality costs correctly among the three categories. The individuals were asked to give estimates of these percentages for each category.

- Itemizing quality cost data—Quality cost data was collected and categorized for each month from May 1975 to May 1976. Appraisal data was determined first, as suggested earlier. Table II details this data for the month of October 1975 only. The time percentage values of Table I were used for all 13 months of data since monthly fluctuations were not considered extreme. Tables III and IV present the October data for internal failures and prevention. Note that the purpose for each cost is the same as listed under the description of each category in the beginning of this paper.

- Percentage breakdown of costs—The percentage of total costs expended for each category was determined for each month. For October 1975 the results are shown in Table V. The percentage values are characteristic of most organizations as discussed previously. Failures account for about 67% and appraisal about 25% of total costs, while prevention is predictably low at 8%. Figure 4 is a plot of cumulative percentage expenditures by category for each month. Also shown are the total quality expenditures for each month. It can be seen that, even though total spending in March went up, the appraisal portion decreased slightly.

Analysis of data/management

The data collection phase has generated some good elementary quality cost figures to be used and manipulated in the detailed analysis. There are several bases that can be used to analyze quality costs. Most of these are covered in publications already referenced, especially [3]. However, since the relation to quality costs and management was the study objective, quality costs were first normalized to production by using the standard cost of production system already established at the plant. This also neutralizes the inflation effect. For each category an index of the quality cost per $1000 of standard cost of production was computed using the formula

$$\text{Category index} = \frac{\text{Quality cost in category}}{\text{Standard cost}/\$1000}$$

For example, in October standard cost of production was $2,250,000. The prevention index is

$$\text{Index} = \frac{3925}{2,250,000/1000} = 1.7$$

The usual quality cost analysis would proceed with a study of graphs and figures similar to Figures 4 and 5. We can see that prevention spending has continued at about the same level and appraisal has grown steadily. Figure 5 indicates that the failure index decreased dramatically and then began a steady ascent in 1976. This means that

Purpose	Function	Percent of time	Monthly cost	Monthly total
Scrap	Supervision	40	$ 220	
	2 Operators	100	400	
	Reports	—	14,100	$14,720
Repair and Rework	Supervision	60	$ 330	
	3 Operators	100	8,600	
	Utilities	—	275	
	Overhead	—	275	9,480
Failure Analysis	32 Technicians	100	$9,920	9,920
Total				$34,120

Table III. Summary of internal failure costs for October, 1975.

Purpose	Function	Percent of time	Monthly cost	Monthly total
Quality Planning	Manager	30	$480	
	Asst. Manager	40	320	$ 800
Process QC	2 QC Supervisors	35	$385	385
Equipment Design	Eng. Manager	20	$320	
	Test Engineer	60	330	650
Training	Manager	10	$160	
	Asst. Manager	10	80	
	2 QC Supervisors	10	110	350
Other Departments	Design Engr. Mgr.	30	$480	
	2 Design Engrs.	30	330	810
Misc. Prevention Costs	Manager	15	$240	
	Asst. Manager	20	160	
	Secretary	30	90	
	Analysts	30	440	930
Total				$3,925

Table IV. Summary of prevention costs for October, 1975.

82

entire cost of quality.

Collection of quality costs

Once you have determined what data you need to perform a quality cost analysis, you have to determine how to get it. Accounting records will help you initially. However, the object of a well organized quality cost system is to set up and maintain a set of quality cost centers which will gather and report the desired quality cost data in the future. Typically, the presently maintained records of a company are not arranged to readily reflect quality costs on such items as scrap cost, all types of inspection costs, quality engineering costs, vendor relations costs, etc. It will take a while to determine which records and reports should be kept so that an ongoing quality cost analysis can be maintained.

• Be prepared to be disappointed with the initial results, because of the monumental tasks that must be performed at this plant.

The quality situation presented in this case study makes it relatively easy to determine the quality costs for each category and analyze the problems in the plant. However, it is not the ease or difficulty of quality cost analysis that is important. It is how the results are used to better the quality image of the product.

Industrial engineers working in quality control must strive to make this function strong. Quality control should make management—top and middle—conscious of:

---

the total cost of scrap and rework went down rapidly and then started to climb again. Computation of separate index values for scrap and rework shows that the value of the scrap index (in terms of standard cost of production) went from 39.0 in May 1975 to 4.9 in December while the rework index was relatively constant. In terms of actual quality dollars expended on the scrap function (see Table III) the plant went from $49,950 in May 1975 to $12,200 in December, a 76% decrease! Note however, that the appraisal index steadily increased during this period.

In order to accomplish the objective, it was decided to introduce management personnel changes into the analysis. The times at which pertinent position turnovers or abolition took place are shown below.

January 1975	Plant manager
March 1975	Manager of materials procurement
April 1975	Supervisor, product acceptance
June 1975	Manager, manufacturing
July 1975	Design engineer
August 1975	Manager, manufacturing engineering
September 1975	Design engineer
February 1976	Production superintendent
May 1976	Plant manager

There has been only one quality control position turnover since January 1975—one supervisor of product acceptance. Of course, the large number of personnel changes indicate an organization in constant transition.

At the beginning of this case study you were asked to assume that you are a new manager of a plant which had a history of quality problems and virtually no quality personnel turnover. You are the new manager assuming the position in May 1976 and have requested that a quality cost study be performed.

Analysis of past quality improvement programs shows that the previous plant manager asked quality control to do an analysis to decrease the scrap. The results are shown in the reductions in the failure percentage of total costs, Figure 4, and index, Figure 5. The preventive cost category decreases throughout the study period. This is due to a cost reduction program which resulted in the loss of personnel in design engineering and the removal of the position of manager, manufacturing engineering (August 1975). This required reorganization of production and process control with increased workloads and decreased communication in several areas, including quality assurance.

In the last few months of 1975 a trade of prevention for appraisal costs was made, Figure 5. This management decision came when the effects of the scrap analysis of 9 months earlier were diminishing and product modifications were just introduced on the production line. Rather than troubleshooting the production problems by preventive analysis, the number of inspections was increased so most defects could be found and repaired. Once again the internal failure index started to climb rapidly. This trend continues through the remainder of

Category	Cost	Percentage	Cumulative
Prevention	$ 3,925	7.9%	7.9%
Appraisal	11,725	23.6	31.5
Internal Failure	34,120	68.5	100.0
	$49,770	100.0%	

Table V. Breakdown of quality costs for October, 1975.

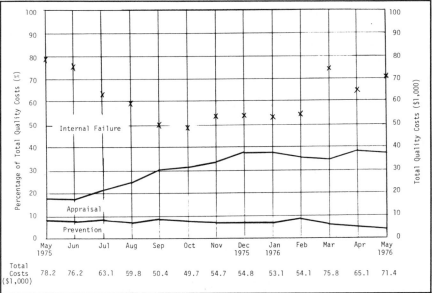

Figure 4. Percentage of quality costs expended in each category.

- The present quality situation in the plant.
- What quality control can do to improve the situation in terms of defect presention. This may mean the initiation of a basic quality cost study.
- How important management is to the quality improvement trends of the plant because of the power over spending. This definitely means that quality control should suggest ways for top management to allow quality control to improve quality.
- The fact that increased defect prevention spending is an effective way to reduce total quality costs.

References

[1] Special Report, "Money (mŭn′ē) n., 1: a measure of value," *Quality Progress*, September, 1975, pp. 24-25.
[2] Juran, J.M. and Gryna, F.M., Jr., "Quality Planning and Analysis," McGraw-Hill Book Company, New York, 1970, pp. 37-69.
[3] ASQC, "Quality Costs—What and How," American Society for Quality Control, Milwaukee, WI, second edition, 1971.
[4] Feigenbaum, A.U., "Total Quality Control," McGraw-Hill Book Company, New York, 1961.
[5] Solorzano, J., "Investigation of Quality Costs for Improved Management," (unpublished MS thesis), University of Texas at El Paso, 1976.

---

the period. Note that none of these trends changed, even with new personnel in design (September) and production (February).

There are other indicators that can be used to analyze the situation; some good in one way, but poor in others. For example, total quality costs decreased through October, Figure 4, and remained fairly steady into February 1976. However, total production was fluctuating from September to March, so the indices of Figure 5 give a better idea of quality cost trends since they are production normalized. Generally, Figure 5 shows a fading preventive program, increased appraisal to try to overcome the diminishing effect of past effective preventive-type analysis. These results are characteristic of a weak quality assurance program, and a plant management attitude which believes that tighter control means more measuring to detect and repair defects. This generic type of quality philosophy has helped in causing the present large staff of 32 technicians in failure analysis, Table III, and a separate function for salvage operations, Figure 3, which has put quality control in the repair business. This can only increase the tension between production and quality control.

Now the new plant manager (you) must face a very bleak situation—weak quality analysis on a newly modified product, production and quality working against each other, low morale, and a relatively green staff in several key management positions.

Suggested future actions

Naturally, the new plant manager has many things to do—cost reduction, staff development, improvement of plant reputation, quality improvement, morale building, etc. Quality must be considered no less important than any of these other factors. In fact, a success in quality improvement may help some of the other situations. Some of the first actions the new manager should take are listed.

- Recognize that there is experience present in the existing quality control staff. Ask for support, help, and guidance from these people.
- Initiate a special quality study to determine defect sources, training needs, etc. This will initially increase prevention costs.
- Return the salvage operation function to production where it belongs.
- Make special efforts to improve communication between design, production, and quality control; at first through you, then expect continued efforts to improve quality.
- Request that quality control continue the collection, analysis, and reporting of quality cost information.
- Establish and maintain quality objectives for the entire plant and specific appropriate people.
- Sincerely realize that top management guides the quality image of the plant.

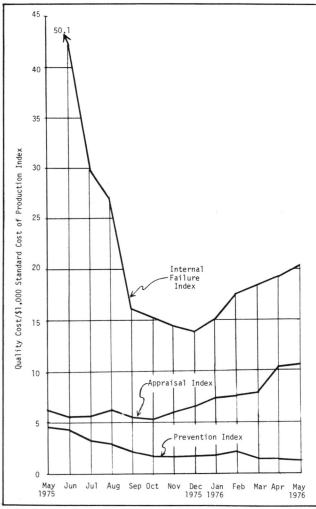

Figure 5. Index value by category of quality cost per $1000 of standard cost of production.

CHAPTER 6

QUALITY MOTIVATION

Shortcomings of Current Motivational Techniques

E.F. Thomas
Manager, Product Assurance
General Dynamics Corporation
Fort Worth, Texas

PREFACE

During the 1976 ASQC Annual Technical Conference, I presented a paper entitled "People - Causes and Effects." In that paper I discussed various motivational techniques and the resulting positive effect on people. Afterward, I suggested to Paul Marti, then Chairman of the ASQC Quality Motivation Technical Committee, that perhaps a paper should be prepared with emphasis on the shortcomings of motivational techniques, so as to provide a balance to the information available to motivation practitioners. Shortly thereafter, Doug Groseclose, current Chairman of the Quality Motivation Technical Committee, called and invited me to prepare a paper along the lines I suggested. I hesitated accepting the invitation because of the potential of such a paper appearing to be too negative. After all, we are prone to suspect those who criticize without offering alternative solutions. I was quick to point out to Doug, as I do here, that there is a universal lack of alternative solutions to the motivational problems we cope with today. Even so, we both agreed that this paper may aid the greater effort to focus attention on our common problems and to develop solutions to the shortcomings of current motivational techniques. Accordingly, this paper will not attempt to suggest solutions; instead, it will point to areas where solutions are considered necessary.

INTRODUCTION

Knowing that motivation influences productivity and job satisfaction and by having much knowledge of what motivates people, one might assume that all American workers at least are totally satisfied and extremely productive in their jobs. However, casual observation will lead one to conclude that this is not true.

During the turn of the century no one was particularly interested in motivation. Instead people were hired, paid a wage, and put to work. This was all based on the assumption that since management paid the wages, people were obligated to do what management told them to do and they were told how to do it. If they failed to do what they were told they were fired and replaced by someone who would do exactly as told. In other words, the human element was reduced to the same status as that of a machine.

Economic and social conditions forced a change in the efforts to motivate employees. As a result employees were motivated by getting them to see the economic aspects of the job. This was done by emphasizing the wages they were paid, the fringe benefits that were offered, clean working conditions, and security, which were provided. The implication was that since management did all of these wonderful things and provided all of these benefits, surely people would want to cooperate and produce. This technique is not much better than the first although it is still widely practiced to date. Although both techniques are still practiced to some degree, it is commonly recognized that the first technique does not work. It is not recognized universally, however, that the second technique does not work.

During the 1930's a new movement in motivation began. Along with economic elements, the social aspects of the job were stressed. This began the era of keeping everyone happy and creating a climate where everyone smiled at everyone else. Did this approach fail to see that some people can be so happy they do not want to work? Possibly.

For 25 years or more, we have made a wide variety of attempts to make the employees feel better about their job and the company. These efforts have resulted in higher wages, broader benefits, better human relations, varied employee activities and management/employee communication programs. These things were recognized to be very necessary and were considered to be welcome additions to the work experience. But lasting motivation among the workers was not achieved.

As it stands today there are not any motivation techniques that could be called current. Most of the motivation techniques that are used were developed from the period from the middle 1920's to the late 1940's or even early 1950's. Such techniques as more money, short breaks in the middle of the day, and shorter working hours are being used today just as they were in the past to motivate workers.

With the above in mind, I would like to examine and comment on some of the motivational techniques widely employed today and point out some of the basic shortcomings of these techniques. This paper will not attempt to explain the techniques themselves. The reader is invited to take advantage of the many thousands of volumes available in any current library which will extol the virtues and the benefits to be derived from proper implementation of these techniques.

MONEY

In researching motivation a common denominator keeps cropping up: money, pay, wages. The opinion of whether it is a motivator or whether it is not a motivator is a function of the last researched report. For example, one writer states that "The monetary rewards can motivate a worker while not satisfying him." In another source we will find "One of the major motivating forces to remain in an unsatisfying job is money." Another states that "Behavioral scientists conclude money is not really a motivator." Frederick Taylor's basic premise was that money is a prime motivator. In standing back and looking at the total information available in current literature, one must conclude that if money is a motivator it is sorely misused in most cases. And if it is not a motivator, it is incorrectly being used as one.

Money is incentive which has the possibility of fulfilling most of our needs. The theory that the harder we work the more money we make holds true in some cases, but not always. The money motive will increase productivity until our material wants become relatively satiated and as the individual becomes confident that he has the ability to earn as much as he needs, the money motive begins a subtle decline. For most people material desires do not multiply ad infinitum, and their appetites will at long last become a bit jaded.

An example of motivation in reverse is the so called "merit raise." Although most companies have some form of a merit raise program, the programs fail because the actual increase in salary between high performance and low performance workers is usually insufficient to stimulate interest in the extra effort required to gain the increase. This leads one to conclude that the claim of behavioral scientists that money is not a motivator may be incorrect. The real issue might be that it is seldom used in a manner that might engender improved performance.

Many in management feel that a raise in salary should be a major motivation for employees. But Rensis Likert found that productivity was consistently highest where both the financial _and_ the psychological needs of employees were satisfied on the job. Therefore, to provide only financial rewards is to create, at best, only partial motivation. Others feel that wages and fringe benefits can only be used as satisfiers, not motivators. The worker cannot spend his wages at work, only after he leaves work. For many wage earners, work is perceived as a form of punishment which is the price that must be paid for various kinds of satisfaction away from the job.

Money incentives will stimulate some to work harder but the individual who has a strong drive to achieve will work hard anyway as long as it is possible for him to achieve. Such a person is interested in money rewards or profit primarily because of the feedback they give him as to how well he is doing. Money is not the incentive to effort but rather the measure of his success.

Most factory monetary incentive plans are unsuccessful because workers do not want to be considered "rate busters" and ostracized or made to feel uncomfortable. Here, group pressure is more important than extra money. Also, once a worker reaches a certain financial level, they look for other life satisfactions which mean more to them than money.

INCENTIVE PLANS

Incentive plans do not take account of several well demonstrated characteristics of behavior in the organization in which they are employed. Incentive pay plans do not consider that most people also want the approval of their fellow workers, and that if necessary, they will forego increased pay to obtain this approval. In addition, no managerial assurances can persuade workers that incentive rates will remain inviolate regardless of how much they produce. Further, the ingenuity of the average worker is sufficient to outwit any system of control yet devised by management. Even though the incentive pay plan may bring about a moderate financial increase, it also may bring a considerable variety of protective behaviors, such as deliberate restriction of output, fudged records, and grievances over rates and standards. It usually creates an attitude which is the opposite of that desired; antagonism toward those who administer the plan.

There are other reasons why workers do not like incentive plans. (1) Workers often feel that if they produce an exceptional quantity of one part, management will cut the rate for that part. (2) Workers feel that employment will be reduced if the work force produces too much at one time, leaving nothing for the next day or the remainder of the week. (3) Uncertainty arises relative to social relationships if one or several dare exceed the groups production quota and outshine others in the group. (4) Some workers feel that incentive plans tend to eliminate the satisfaction derived from maintaining at least a minimum amount of control over their own behavior.

In providing rewards management must be able to provide commensurate rewards. An employee may know that he is regarded as one of the best employees, but this will do little to enhance his performance if his supervisors cannot provide him with sufficient rewards or opportunities to gain rewards. Further, the reward has little effect unless it is public. The fact that a person's peers know about the reward makes the effect much greater. In fact, this may mean more to him than the reward itself. Knowing this, it is surprising that many companies still attempt to keep secret the size of merit increases.

PUNISHMENT

Man's earliest motivational technique was probably the whip. Today this technique is still widely practiced. Punishment or a threat of punishment is believed by many supervisors to be an effective device for securing obedience from employees and it may occasionally have some value. The negative approach too often is used because it is usually easier to punish the employee than to provide positive guidance and assistance. Punishment, however, often introduces other factors that make it considerably less desirable than positive forms of motivation. For example, punishment is likely to frustrate the individual being punished with the result that he may become aggressive or perhaps childish in behavior. These reactions usually make it more difficult for him to change his behavior in the desired direction. The threat of punishment may also cause the employee to become sufficiently fearful that he is not able to respond in a positive manner to suggestions and instructions. He is likely to develop unfavorable attitudes toward everything associated with the job. At best, under the circumstances, he seldom performs with enthusiasm. Further, it is quite likely to create resentment because if the disciplined worker believes, and especially if the work group believes, that the boss has been unfair, cooperation may be greatly decreased and may well disappear altogether.

Rewards and punishment are easily equated to the "carrot and stick" theory whereby people are seen as jackasses with management having a stick in one hand and a carrot in the other. The characteristics of a jackass are stubbornness, willfulness, and unwillingness to go where someone is driving him. These, by an interesting coincidence, are also the characteristics of the unmotivated employee.

PRODUCTIVITY VERSUS JOB SATISFACTION

Many people in the field of motivation equate productivity with job satisfaction. The general assumption is that as long as the individual is personally satisfied with his work, motivation will be high and directed toward high output. In other words, management seems to assume that a person can be induced to increase work output out of a feeling of gratitude to the company or to the work system.

All too often the motivational technique of attempting to increase job satisfaction to achieve greater productivity simply does not work. Personal satisfaction can be more nearly equated to achievement of personal goals and objectives. If one's personal goals and objectives are to earn just enough money to support oneself in a certain manner or style of living, once that amount of money is being earned efforts to increase job satisfaction/productivity will be ineffective. Further, those people who are members of informal work groups in which group norms have been established and enforced will generally turn a deaf ear to all attempts to motivate to higher levels of productivity. Workers have been known to say that the reason they work four days a week is that they cannot earn enough money to support themselves by working three days a week. Obviously people who think in this manner are beyond almost any motivational approach.

PARTICIPATIVE MANAGEMENT

Participative management relates to a classless cooperation between workers and managers in an attempt to harmonize relations between the two groups and develop and increase production output.

Shortcomings of participative management might include the movement away from the "I" of management to emphasis on the "we." This idea leaves many ambitious managers disgruntled about their lack of personal recognition. They feel that they contribute to the effectiveness of participative management but receive no special recognition for above-average performance. Further, it seems that this thought seems to demotivate some managers who were initially intended to be motivated themselves.

Another drawback to participative management is in the area of communications which is supposedly one of the strengths of the technique. However, if management is not careful, they will be prone to having too many people in far too many meetings. It will seem that nothing can be done without the team management concept applied. This delay caused through communications can also cause a delay in production.

The union influence still remains strong. The blue collar worker's union still clings to the tenets of separation of management and labor. The labor unions seem relentless in not releasing their control.

One risk to participative management is that many employees may not fully understand the objectives and goals of the company. Therefore, their decisions may tend to lean toward their own personal feelings. These decisions can have a definite impact on the total organization. Therefore, they should be made by people who are as capable and responsible, i.e., without bias, as possible. Finally, employees may tend to be biased toward ideas and feelings of their own informal work groups when making these decisions in a participative management system. This can be a major problem, especially if the employee's informal work group goals do not parallel with the company's.

JOB ENRICHMENT

Job enrichment is a program designed to reshape and redesign particular jobs in an effort to make the jobs more enjoyable, and hopefully more productive.

The major drawback to job enrichment is the attitude managers have in relation to it. Many managers feel that this system is a cure for all ills at once. In some instances job enrichment attempts might add even more complexity to an already chaotic situation. Many managers feel insecure in their ability to deal with the kinds of problems that inevitably arise when a major organizational change is made. Top management has often failed to assess its own commitment to job enrichment. An inability to cope with change will reflect disaster in any job redesigned project.

In one situation there was a massive redesign of jobs. An evaluation afterwards revealed an interesting situation. It turned out that people who held good jobs before the change also held them afterwards, and the people with repetitive and routine positions had essentially identical jobs after the work was redesigned. All that had actually happened was that the office had been physically rearranged and the names of some jobs altered, but the jobs themselves had remained virtually untouched. In several cases, deterioration of morale, increase in tension, and resentment had developed through the backfire of job enrichment.

In a job enrichment program supervisors are often prone to assign most of the difficult tasks to their key employees and to assign the dull jobs to those employees who are marginal or weak. This can be defeating in the long run. What must be done is to provide opportunities to find challenging and interesting work to all employees.

In general, there are six major things that can go wrong in organization redesigned work. (1) Rarely are the problems in the work system diagnosed before the jobs are redesigned. (2) Sometimes the work itself is not actually changed. (3) Even when the work is substantially changed, anticipated gains are sometimes diminished or reversed because of unexpected effects on the surrounding work systems. (4) Rarely are the work redesign projects systematically evaluated. (5) Line managers, consulting staff members, and union officers do not attain appropriate education in the theory, strategy, and tactics of work redesign. (6) Traditional bureaucratic practice creeps into work design activities.

PAT ON THE BACK

One motivation technique that is one of the most important is simply telling an employee that he is doing a good job. To have the boss come to your office and sit down with you and tell you that he appreciates the hard work is one of the very best motivating factors. This makes a person feel needed and responsible. This technique can also be very harmful if not used correctly. For example, the boss may have a favorite employee and show more appreciation to him or her and make the other employees jealous of the attention. Other employees will feel they work just as hard but may not get the personal recognition which is greatly needed in almost everyone. This is a technique which, when used in the right manner, can do much good; but when used improperly, can give a devastating blow to the work force.

BEING ONE OF THE GUYS

Sometimes in an effort to motivate their people managers will play down their authority and play up their likability by attempting to act out the role of a "nice guy." This is referred to as status stripping, where the individual tries in a variety of ways to discard all the symbols of his status and authority. Such efforts range from the open door policy to democratizing work by proclaiming equality of knowledge, experience and position. Much to his horror the manager finds that attempts to remove social distance in the interest of "being one of the guys" has resulted in reduced work effectiveness. He discovers that his subordinates gradually come to harbor deep and unspoken feelings of contempt toward him because he has inadvertently provided them with a negative picture of the rewards that await them for achievement, i.e., a supervisor who is nothing more than "one of the guys", a prospect unpleasant to contemplate.

PERFORMANCE APPRAISALS

In many companies the performance appraisal has proven extremely effective in motivating people to higher productivity levels. A primary factor in the success or failure of this technique is the supervisor himself. In practice, some supervisors find it very difficult to appraise the performance of others accurately and sometimes find that it is particularly difficult to tell others the negative things about themselves. Therefore, appraisals can lose their effectiveness and even their validity if they are not administered both honestly and thoughtfully.

The reaction of the employee can become very predominant in the performance appraisal process. If the employee is dissatisfied with the performance, supervision, or even the personal traits of the supervisor, any praise or criticism from the supervisor will carry little or no weight. If this be true, any benefits from the performance appraisal will be negligible.

Lack of motivation is not the only factor involved in poor performance. An employee may be highly motivated but certain barriers prevent good performance. Such barriers, to name a few, include lack of adequate authority, a poorly defined job, an ineffective interaction influence system, poorly defined goals and faulty strategies for achieving them, inability of supervision to clearly and adequately relate what they really mean when counseling workers during a performance appraisal, poorly constructed performance appraisal procedures, and inadequate documentation of worker performance.

Cases have been reported which reveal a large percentage of the employees are clustered into predetermined categories, ranging from below-average to above-average performance. In one particular company there existed pressures to limit the number of better than average ratings so as not to upset budget allotments for salary increases.

Many managers feel that they are not in a position to objectively judge others. There are few effective established mechanisms to cope with the feelings of inadequacy managers have about appraising subordinates, or the paralysis or procrastination that results from their guilt about playing god. Because of the many problems associated with working with appraisals and analyzing the data obtained from the appraisals, many managers fall into a very difficult situation of timing. Although managers are urged to give performance feedback freely and often, there are no built-in mechanisms for insuring that they do so. Delay in feedback creates both frustration when good performance is not quickly recognized, and anger when judgment is rendered from inadequacies long past.

Another problem area of performance appraisals is focused toward the inconsistency of comparisons. Ratings by different managers, and especially those in different units, are generally incomparable. This situation is caused by different personal interpretations of performance criteria and by varying degrees of communication gaps between different supervisors and their people.

GOAL SETTING

The technique of goal setting is probably one of the most powerful motivating techniques of modern management. However, inexperienced supervisors fail to motivate workers because they are afraid to demand enough. Even experienced managers do not realize that the establishment of goals is a strong psychological and motivating technique.

There are pitfalls to be avoided. If the goal is too difficult it will have little or no motivating force. Some disagree with this, much to their regret. They say, demand twice what is expected and people will automatically put forth extra effort. This approach may work once or twice, but the long range consequences are severe. First, people become frustrated because they can never reach the goals which are set. Second, there can be no sense of achievement and/or accomplishment. Third, people are constantly on the defensive in terms of justifying actual levels of performance. Fourth, padding of budgets and schedules takes place. And finally, in time, the superior-subordinate relationship becomes one of antagonism.

HUMAN RELATIONS THEORY

Under the human relations theory every employee is to be provided the right to influence, if not to actually choose, how they will perform their assigned tasks. In large companies the implementation of this theory is often quite difficult, especially if the manager is not inclined to establish this type of working relationship with his subordinates. Further, insofar as management is concerned, there is no recognized need to structure every job to provide an opportunity for personal growth and need satisfaction. Many people in management recognize that a business or an industry cannot efficiently accomplish its goals and objectives when every department is run with due regard for every ego in it. This is why human relations, when used as a motivational tactic, is often ineffective.

DELEGATION

The objective of the delegation technique of motivation is to get the employee to make his own decisions, although management still has the final word. The purpose of this technique or objective is to instill in the employee a sense of having a greater degree of control of his destiny in the work environment, thereby deriving a sense of personal satisfaction and perhaps even satisfying his ego needs. Some feel that such an approach will inevitably lead to higher levels of productivity.

In spite of the above the delegation technique of motivation can cause problems because an employee could make a decision in which he had contributed a lot of time and thought and then management could reject it. When this happens the employee can be more unhappy than he was before he was given the delegated task.

Further, many executives fail to observe the principle of delegation, perhaps because of a fear that the subordinates may threaten their status or job security, or perhaps because these executives do not want to admit that subordinates might be able to perform their duties equally well. Superiors at times also may be unable to delegate authority because they have failed to develop subordinates who can assume authority. For example, an employee may possess the requisite ability to make decisions, but may not have sufficient confidence. Because of this lack of confidence, the delegation technique of motivation may very well prove disastrous, for some people are afraid to be turned loose to succeed or fail.

WORKING CONDITIONS

Some people in management consider a positive change in the working environment to be a motivation technique. Such a change would include modernizing and maintaining the equipment which the workers use on a day to day basis. Also included are improvements in the physical condition within the office or plant. However, there is some disagreement as to the usefulness of this technique. In general, in most working places today, the physical conditions are good. Temperature and ventilation are adequate. Rest periods and periodic coffee breaks are provided. Lighting is good and most working needs are satisfied. Thus, the presence of good working conditions in most plants and offices today is taken for granted and have little, if any, motivating force. Further, these adequate conditions are usually guaranteed by laws or by the union rules, and therefore even the insecurity of having these benefits removed cannot in itself act as a motivation for higher productivity or for better quality work.

SURVEYS

Sometimes management decides to determine the attitudes of the employees so as to construct proper working conditions, working environment, etc. To make this determination, attitude surveys are developed and distributed to each employee. The image projected by management is that if the employees will only indicate what will make them happy, management will be only too glad to oblige. The employees are encouraged to frankly and candidly state their complaints relative to management policy and shortcomings. Questions are worded so that no punches need to be pulled by the employees. There is no requirement for the employees to reveal their identity.

Efforts to solicit inputs from the employees all too often fail since the employees have observed from past experience that management too often pays too little attention to the results of the surveys. When management allows individuals to state their thoughts only as a source of satisfaction to the individual to "get things off his chest", then the potential benefit of future surveys is greatly diminished. Management should never attempt an employee attitude survey without a predetermined commitment to "do something" with the survey results.

UNIONS

Many jobs found in manufacturing operations are represented by unions. Even though survey after survey indicates that money is not considered to be as important as many other things, such as working conditions, opportunities, potential for growth, etc.; whenever union representatives and management get together across the negotiating table, the primary point in negotiations is "money." As Abraham Maslow suggests in his need hierarchy, money is the primary satisfier for physiological and safety needs. Many will agree with Maslow that once these needs have been satisfied, there is inevitably a shift to social and perhaps self-esteem needs. Yet unions and management still focus their attention to more money, i.e., greater provision to satisfy physiological and safety needs.

Unless there are opportunities in the work place to satisfy higher level needs these needs will in general be not satisfied and employee behavior will reflect this deprivation. Under such conditions motivational efforts on the part of management are bound to be ineffective.

YOUNG PEOPLE

A prime reason for motivational troubles is that all too many American workers - particularly young ones who are supposed to be bubbling with energy and ambition - no longer care. This has resulted in absenteeism rates as high as 20 percent on Fridays and Mondays in some manufacturing plants, forcing management to rustle up part-time student help to keep the assembly lines going. This problem can be traced to two main factors; a younger work force, 25 percent of which is under 25 years old, and the nature of the work itself in a highly industrialized society.

The young workers today want more satisfying jobs, greater participation in decisions affecting their working lives, and an end to authoritarianism. The elevation of living standards has led young workers to believe that they are entitled to benefits such as secure retirement, a meaningful job, and health care coverage as a matter of social rights. In order to satisfy the young workers management will have to change with them and meet their needs, or else lose them as effective workers. Work for the sake of work holds little value for younger workers if the job is not satisfying.

THE JOB ITSELF

Some jobs have not changed to coincide with the different wants and needs of the workers. There is a difference between what people are looking for in a job today and what they sought a generation ago. A decent living and job security were paramount in earlier days. But today, certainly, expectations go beyond these. Many of the jobs that still exist in industry were designed for an earlier period in American history. It was a time when you had an influx of immigrant labor, willing to work for little money at repetitive, mind-dulling tasks. In short, jobs were engineered for replaceable workers. Due to the technical skills required for many of today's jobs, along with a form of organization of labor unions, the workers are now able to demand more of management.

Today, many employees may find their social and egotistic needs satisfied away from the job and may work only to be paid to satisfy their physiological needs. The job itself holds no interest and is merely a means to an end. These people are active in the union, or church, or a hobby club, and fulfill their social and egotistic needs in that manner. For these people, there is likely to be a minimal relationship between need satisfaction and job performance.

Some feel that workers could be motivated by providing them a sense of job security. Certainly job security is an important internal force affecting the worker. Job security can influence the loyalty felt toward an organization, which should in turn have some impact upon the productivity of the worker. However, if the worker becomes too comfortable with the security of the job, a sense of laxity may result. On the other hand, in a low job security environment, morale is generally very low. There is little or no loyalty felt toward the organization and turnover is extremely high.

RESULTS OF POOR MOTIVATION

A costly problem for industry during extended periods of economic prosperity has been the high turnover of employees. The fact that many jobs are interchangeable among companies facilitates job switching. People who are not properly motivated generally tend to exhibit high degrees of job dissatisfaction, which probably explains why "quit rates" in some industries jumped 80 percent from 1964 to 1973. People will tend to leave situations in which they are dissatisfied or do not derive satisfaction. High turnover rates are costly in terms of the direct expenses involved in hiring and training of new personnel, and also in terms of the hidden costs of supervisory coaching and break-in time given by other employees. Further, if the people who leave the company find better opportunities elsewhere, valued employees with longer service who might otherwise stay with the company are encouraged to take the plunge into the job market. This is either a case of dissatisfaction breeds dissatisfaction or success in greener pastures breeds a desire to share in that success.

Job dissatisfaction certainly contributes to absenteeism. If people are unwilling or unable to quit a job, one way that they can show their displeasure is by staying away from the job as much as they can. Closely related to absenteeism is the problem of tardiness. This too has been on the increase with the result that production lines have been prevented from starting up promptly when the shifts change in manufacturing plants, and telephones and desks go unattended in offices. Another spinoff of absenteeism is leaving the plant early. If you have ever gambled with your life by standing in a plant parking lot at quitting time, you know just how anxious people seem to be to get away from their work place by the manner in which they roar out of the parking lot.

MOTIVATION ITSELF

When exposed to various motivational techniques some workers are left with the impression that they are believed to be so unintelligent and naive that motivational techniques can be used on them effectively, even repeatedly. When this is perceived the worker is likely to rebel, inwardly if not outwardly. And why should this response be unexpected. After all, the workers have been to the same schools as the motivators and taken the same motivational courses. Surely the workers can recognize most motivational techniques almost instantly. How long has it been since a new technique has been developed.

Herein lies the danger of attempting to motivate people in a clumsy manner. It downgrades those on whom the motivational technique is practiced. And when the worker understands what is happening, the motivational techniques are perceived as tricks, which not only fail to work but give rise to resentment against anyone who attempts to use such techniques. Many managers make this mistake and are completely in the dark as to why the workers follow without enthusiasm or follow not at all.

THE PETER PRINCIPLE

With tongue in cheek I would like to point out that according to the Peter Principle, in time every managerial position tends to be occupied by a person who is incompetent to carry out its duties. To the degree that this is true, the upper levels of management tend to be occupied by incompetent people. To the degree that this is true, efforts to motivate the incompetent will probably and inevitably prove to be futile. Motivational techniques today presume competency on the part of the ones being motivated.

CONCLUSION

Is it possible that we have been laboring under a misconception by trying to motivate people? It is an interesting phenomenon that after many years of experience with efforts to motivate people, there is remarkably little tangible evidence to indicate that people are any more or less motivated now than they were before.

In closing, one question we might consider is whether or not people really need to be motivated. From a psychological standpoint, men are motivated from birth. They are striving, seeking, purposeful creatures. Psychologists also suggest that work itself may be an instinct of man.

At the beginning of this paper I stated that solutions to our motivational problems would not be recommended within the paper. Yet, I cannot resist a desire to offer an approach to arrive at a solution which can be universally applied. To establish the framework for making the suggestion, many people currently feel that lack of motivation is merely the result of an incompatibility of employee objectives and organizational objectives. A newer view of this conflict of objectives is developing which acknowledges that employees may well enter an organization with objectives different from those of the organization itself. Realizing this, managers must stop complaining about the fact that the objectives are different. Managers must accept a new managerial role for themselves. In this role they must accept the challenge to seek ways to establish within the work environment jobs which will enable employees to satisfy their personal goals and at the same time satisfy organizational goals. Instead of wishing that your employees were more like those of a bygone depression era, it might be better to assume that your employees are indeed like you. Reflecting on what motivates and satisfies you, you then might be able to determine those characteristics of your employee's world that are diminished or entirely lacking in your world. Whatever characteristics these may be, if they were made available to your employees, would your employees then be more motivated and productive? Probably.

BIBLIOGRAPHY

Although this paper is not footnoted the author freely acknowledges many references and excerpts from the following sources which comprise much of what we know about the shortcomings of current motivational techniques.

Chruden, Herbert J. and Arthur W. Sherman, Jr. Personnel Management. Dallas, Texas: Southwestern Publishing Co., 1968.

Dunn, J. D.; Stephens, Elvis; and J. Rolland Kelley. Management Essentials. 3rd Edition. McGraw-Hill Book Co.

Galenson, Walter. A Primer on Employment and Wages. Vol. 328. New York, New York: Random House, Inc., 1966.

Gellerman, Saul W. Motivation and Productivity. American Management Association, Inc., 1963.

Giblin, Edward. "Motivating Employees: A Closer Look." Personnel Journal. Vol. 55, No. 2. Santa Monica, California: Arthur Young & Co., February 1976.

Goble, Frank. Excellence in Leadership. American Management Association, Inc., 1972.

Hackman, J. Richard. "Is Job Enrichment Just A Fad?" Harvard Business Review. Vol. 53, No. 5. Boston, Mass.: 1975.

Haimann, Hilgert. Supervision Concepts and Practices. Cincinnati, Ohio: Southwestern Publishing Co., 1972.

Hepner, Harry W. and Frederick B. Pettengill. Perceptive Management and Supervision. 2nd Edition. Englewood Cliffs, New Jersey: Prentice-Hall, Inc., 1971.

Levinson, Harry. "Appraisal of What Performance?" Harvard Business Review. Vol. 53, No. 4. Boston, Mass.

Lewis, Ralph (Ed.). Harvard Business Review. Vol. 49. Boston, Mass.: Graduate School of Business, Harvard University, July 1971.

Marrow, Alfred J. The Failure of Success. New York: AMACOM, 1972.

McGregor, Douglas. *The Human Side of Enterprise*. 3rd Edition. New York, New York: McGraw-Hill Book Co., Inc., 1960.

McQuaig, Jack H. *How to Motivate Men*. New York, New York: Frederick Fell, Inc., 1967.

Morris, John O. *Make Yourself Clear*. New York, New York: McGraw-Hill Book Co., 1972.

Patz, Alan L. "Performance Appraisal: Useful but Is Still Resisted." *Harvard Business Review*. Vol. 53, No. 3. Boston, Mass.

Peter, Dr. Laurence J. and Raymond Hull. *The Peter Principle*. 16th Edition. New York, New York: William Morrow & Co., Inc., 1969.

Sartain, Aaron Quinn, et. al. *Psychology: Understanding Human Behavior*. 3rd Edition. New York, New York: McGraw-Hill Book Co., 1967.

Scanlan. Burt. *Results: Management in Action*. 2nd Printing. Burlington, Mass.: Management Center of Cambridge, 1969.

Steinberg, Rafael. *Human Behavior, Man and the Organization*. New York, New York: Time-Life Books, 1975.

Sutermeister, Robert A. *People and Productivity*. 2nd Edition. New York, New York: McGraw-Hill Book Co., 1969.

"Too Many U.S. Workers No Longer Give A Damn." *Newsweek*. Vol. 79, April 24, 1972.

U.S. News and World Report. Vol. LXXXI, No. 13. Washington, D.C.: News Department and Executive Offices, September 27, 1976.

Weingarden, Jaala. "What Job Attitudes Tell About Motivation." *Harvard Business Review*. Boston, Mass.: 1968.

Zaleznick, Abraham. "Human Dilemmas of Leadership." *Harvard Business Review*. Boston, Mass.: 1963.

Integration Of TQC And Motivation Programs

Dr. B. Veen

Akzo Research Laboratories Arnhem—The Netherlands

SUMMARY

In the past several ideas about job satisfaction have been developed and tried out (job enlargement, job enrichment, etc.). In the long run the experiments proved unsuccessful because they did not contain a built-in mechanism to guarantee growth and continuity. Quality Control needs motivated workers and the solution of many smaller problems. Motivation can be improved by participation in problem solving. QC can provide sufficient suitable problems. So, it was a logical step to try to integrate a TQC program with a motivation program (cf. QC-circles in Japan).
We have made several experiments in this direction and learned from it that they are successful in both respects (QC and motivation).

MOTIVATION PROGRAMS

In the past decades several ideas have been developed about the way to improve motivation and job satisfaction among workers. To mention a few: job enrichment, job enlargement, work consultation, self control, participation, etc. All these methods have in common, that they try to give a greater challenge to the worker. Some try to break monotony, others aim at recognition and self-esteem or at giving more responsibility, while we are told that the primary goal must be purely humane. We should strive for a greater happiness of the worker in his job. We should not consider this as a way to improve economic results. That would be manipulation!
In my opinion this is a misapprehension of the worker's common sense. Every worker understands that a company is not a charity organization. They are even inclined to mistrust such measures, if it is not clear to them that the measures are taken to improve the totality of the company, in which both human and economic aspects are involved. This has been a weak point in the start of many of the earlier experiments. However, there are other reasons why many experiments did not fulfil the high expectations; most of the experiments faded out within a few years.
We tried to analyse these unsuccessful experiments so as to gain a better insight into the requirements for a greater chance of permanent success. We think that the earlier experiments to enhance job satisfaction lacked at least one of the following requirements:

1. Honest humane intentions should be in balance with rational economic goals. Otherwise, the workers will not believe us.
2. A proper introduction and acceptance on all levels must be achieved. The argument speaks for itself. Nevertheless, we have often disregarded this ourselves, because we were too optimistic and too eager to see results.
3. Supervisors should get a careful training in participative leadership. We shall not forget that in the past we selected our supervisors for quite different abilities.
4. Complete organizational units or departments have to be involved. If only parts of such natural groups are involved, one risks feelings of envy or special status.
5. There should be a constant stream of sufficiently challenging tasks and problems. The goals should be attainable but not too easy. This point is very important. It seems to us that many experiments to enhance job satisfaction have failed because of a kind of "Hawthorn" effect. Initially, the results were promising because there was a change and a new challenge, but soon the new tasks became routine again and the motivational aspects died out.

THE INTERACTION OF QUALITY CONTROL AND MOTIVATION

It is well known that quality control needs motivated people to get optimal results. Therefore, QC-managers have always had a real interest in activities which could promote motivation among workers. They have always remained alert to contribute from their own specific field of activities. I can mention, for instance, the principle of self-control. Now again I feel that there is an opportunity for quality control to contribute.
We are faced with a multitude of small QC-problems and by tradition we think that they should be solved by staff people or specialists. However, as we have only a limited number of such people, most of the smaller problems can seldom be studied and solved in time. We even have got a situation where the specialists are constantly busy to "extinguish fires" which have been caused by all these small problems, so that they do

not come to their real task of solving the bigger problems and thinking about future improvements to make the house more "fireproof". A breakthrough is necessary here. And what is more logical than to have the large number of small problems solved by people who are also large in number, viz. the workers on the shop floor.

In Japan this step was taken more than 15 years ago, and it has been found to work. It has also contributed to the motivation of the workers there. In this respect I would underline the conclusion that the Japanese worker is better motivated than most of the workers in our western countries because of his share in more challenging creative brainwork in quality circles, where he can solve problems and gets recognition. We mostly hear the reverse, viz. that he is motivated by himself and is therefore inclined to do a good job and participate in QC-circles. This is not true. To illustrate this, one may ask the categorical question "What about the quality of Japanese products and the motivation of the workers in Japan before the start of the QC-circles in 1959?". Another important feature we could observe in Japan is the high level of general education of the shop floor workers. There are even plants where a secondary school education is the minimum requirement for a permanent job. In dealing with such a high level of education of the workers, management had to find a compensation. The lack of creative work inherent in production line jobs, could not satisfy the needs and ambitions of the workers. Therefore management had to find a new balance between manual routine work and participation in creative thinking.

At the same time the high level of education gave them a favourable starting position, because it simplified the training program needed for more participation in quality problem solving. Nevertheless, there still remains a gigantic task, which is certainly not neglected by Japanese management.

If we now analyse the situation in our western countries, we see that a similar development is taking place. Compulsory education is extended to 9 years or more. People have got a broader general knowledge and insight, also because of the influence of modern communication means. They want to use it. They want more participation in organizing and problem-solving tasks concerning their own work. They want more recognition of their abilities and personalities than they can get from being an extension of a production machine.

In this way we come to a model shown in Fig. 1.

Experience has taught us that it is not possible to change suddenly to a situation in which there is more room for participation in organizational or decision making responsibilities. We shall have to build up experience in this field and create a new climate. This will also call for a new style of leadership, which we cannot expect to change in the turn of a hand. However, we have seen from several experiments that such a change is possible, and that the best way is to start with participative problem solving groups. But this step is not as simple as it looks. If a group of workers is put together and given a problem to be solved, hardly anything will come out. We first have to train group leaders in at least three abilities.

1. How to lead a group discussion.
2. How to solve a problem in a systematic way.
3. How to organize continuation and follow-up.

We feel that training in the first subject is beyond our scope. It should be carried out under the supervision of personnel development people. This guidance should be extended after the first formal training period until sufficient experience has been gained. The second item is clearly a matter of quality control. Here too, it will be necessary to continue guidance during the first period after the formal training. The third aspect is a matter of practical guidance by bosses and the quality department.

AN INTEGRATED PROGRAM FOR QUALITY CONTROL AND MOTIVATION

Three years ago one of the divisions of our company decided to intensify its quality control. This division comprises about 30.000 workers in 25 plants. My department was asked to advise on a program and to render assistance during the initial period. A survey of the plan designed by us is given in Fig. 2.

One of the first activities was to train or retrain for each plant a so-called quality coordinator. This man had to promote and implement the total quality control function in general, and to coordinate the different aspects, such as quality cost assessments and analyses, quality control procedures, motivation programs, etc. These coordinators were trained by our department in methodology and management of quality control. During the first year we also gave them as much as possible assistance and advice. From Fig. 2 it can be seen that besides the traditional quality control activities we included the motivational aspects. However, we did not pursue the program before the upper levels were sufficiently trained in this respect.

We prepared an introductory course in quality control and problem solving adapted to the educational level and needs of our workers and laboratory personnel. This course is given locally by the quality coordinators with the assistance of other staff members of the plant. They also fill in the general course with examples from local practice. A survey of the contents of this course is given in Table I.

Management training in QC was mainly achieved by attending external courses, such as the courses of JURAN, PITT, and others, which are regularly given in Rotterdam by the Dutch Quality Control Foundation KDI. Supervisors are trained by the quality coordinators, but use is also made of the KDI facilities.

A number of our plants have now reached a point where the program is in full operation. They have built up so much experience that we can give more detailed advice on what to do and what to avoid in the case of QC-circles.

Fig. 3 gives a picture of the quality circle and the contacts with its surroundings. The quality circle itself should be composed of people from one department and one shift. The group should comprise not more than 8 and not less than 4 persons. One of the members should be appointed as a leader and he should have sufficient authority. Therefore, mostly a foreman or a chargehand is chosen. He also receives some extra training in problem solving methodology and techniques. One of the other members is made secretary. He is responsible for minutes of the meetings and distribution of written information among the group members. In general, he gets a strong support from the coach. For this work we appointed and trained coaches to assist the group leaders and secretaries; they are chosen from lower management. Their rôle is very important at least in the first few years of the program when the group leaders have to build up their experience. They can manage about 5 teams. Their task is listed as follows:

1. to promote progress of the work of the quality group;
2. to assist in contacting people from other departments, if their advice or cooperation is necessary;
3. to advise on editing, and distribution of minutes;
4. to check completeness, attendance, and regularity of meetings;
5. to guide and watch the application of certain principles like the "Deming circle" (plan-do-check-action) and the "problem-solving cycle" (collect information- possible causes- root causes- possible solutions- best solution- practical proposal);
6. to check follow-up: "ringing the bell" if necessary;
7. to coach the group leader to become more independent;
8. to take care that the work and the results are given sufficient publicity in the company papers, etc.

Problems
　　The problem for a quality circle can be chosen from two lists:
a) the priority list derived from the Technical Improvement program;
b) a list composed by the group itself.
However, in both cases the following restrictions apply:
1. it should have the consent of management;
2. it should be a problem relating to one's own work, product, machines or procedures;
3. it should relate to technical improvement;
4. it should be not too sophisticated.

Minutes
1) A list of conclusions and arrangements is sufficient.
2) Relevant data necessary for subsequent discussions can be added.
3) A cause and effect diagram of the problem should be kept up to date.
4) The minutes should circulate among staff and line for advice or critical remarks.

Proposals
　　A formal final report with proposals will be made for management. Management should react as promptly as possible. If the reaction is negative or the implementation is delayed, the group has to be informed of the reasons why. When the improvement has been implemented, the group should be informed about the results.

Support from management and specialists
　　No motivation program will survive if management is not fully supporting it. This must not only be shown by words, but also by organizational assistance:
1. All members of the group always get the opportunity to come to the regular meetings.
2. Sufficient meeting facilities should be provided.
3. The group should get support for documentation.
All line and staff departments should advise the groups as much as possible and not belittle activities or - which even happens - take an attitude of "you are fishing in my pond". They should use simple language and avoid professional jargon.

Quality circles - rewards and suggestion box

Invariably the question arises if quality group members should be rewarded for profitable proposals, especially in those cases where a suggestion box system exists. There are several reasons why we are not in favour of such a system. The aim of combining TQC and motivation is based on economic as well as humane considerations.
Participation in thinking about one's own work gives more satisfaction, which is a reward in itself. That it gives also better economic results is necessary to be able to carry on in this way, while the surplus is for the benefit of us all. This participation may become a normal way of life and should not be additionally rewarded.
We also have to keep in mind that:
- a quality group works on a problem only after the authorization of management;
- the company provides all kind of facilities;
- the group receives assistance from other people (e.g. specialists);
- the problems are very diversified and the work not only aims at improving economic results but often also the general working conditions.

Nevertheless, when a group has finished its job and is dissolved, the members may get a small souvenir such as a ball point with an inscription, or a certificate.

CONCLUSION

As I have said before, it is too early to claim a complete success of this method. The results, however, have been promising so far and indicate that we are on the right way. We can observe better motivation as well as a lot of solved problems and economic improvements. To underline this statement I should like to end my lecture with some quotations from the report on a visit to Europe by a Japanese Quality Control Study Team under the guidance of Professor Ishikawa, who also visited one of our Enka plants.

"The QC activities at Enka have been founded and built up enthousiastically. The effort and diligence by which the contemporary level has been reached is to be praised highly."
"We believe that Enka as to quality circle activities at this moment is one of the most advanced companies in Europe."
"We could notice a warmth in human relations."
"..... that this company has a well-balanced policy whereby the quality control and the human aspects are equally important. The action to incorporate lowering of the quality costs in the motivation program is an example of it."

TABLE I -- Contents course "Introduction to quality control"

Chapter	
1	Introduction
2	Tables and graphs
3	Motivation and job-restructuring
4	Quality circles and group work consultation
5	Distribution, mean and variance (spread)
6	Making a workschedule
7	Process-control - control charts
8	Quality costs
9	Collecting information
10	Sampling techniques
11	Process descriptions for quality control
12	Problem handling (according to Kepner and Tregoe)
13	Pareto-analysis
14	Traceability
15	Experimental designs

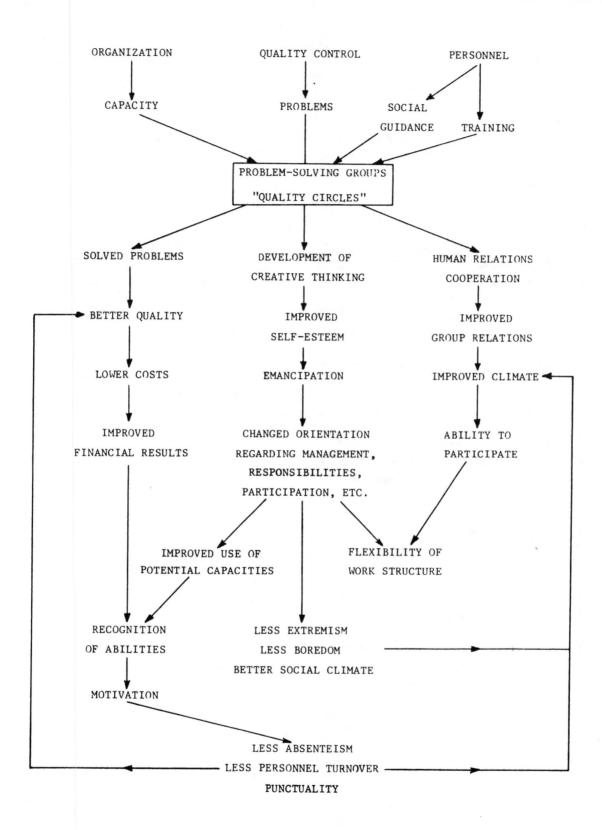

FIGURE 1 -- Consequences of Quality Problem Solving Groups

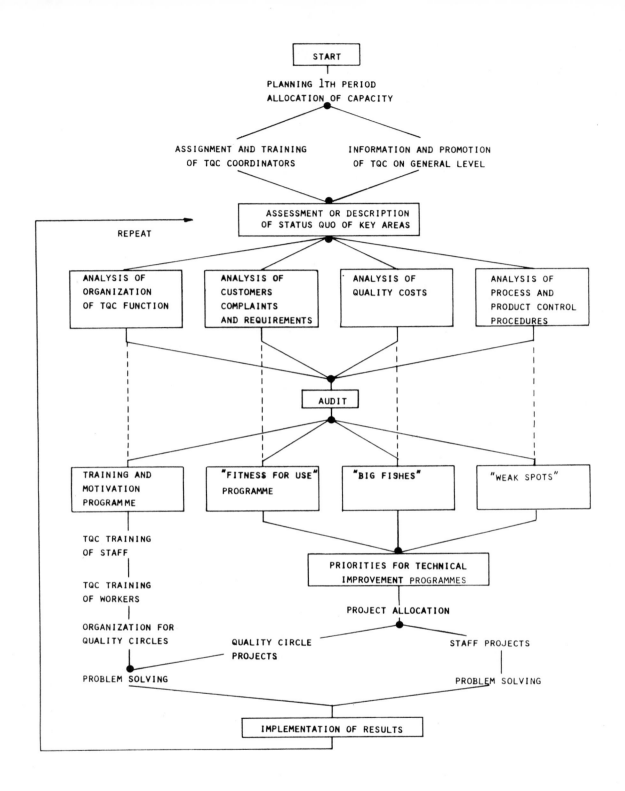

FIGURE 2 -- Integrated TQC-Motivation Program

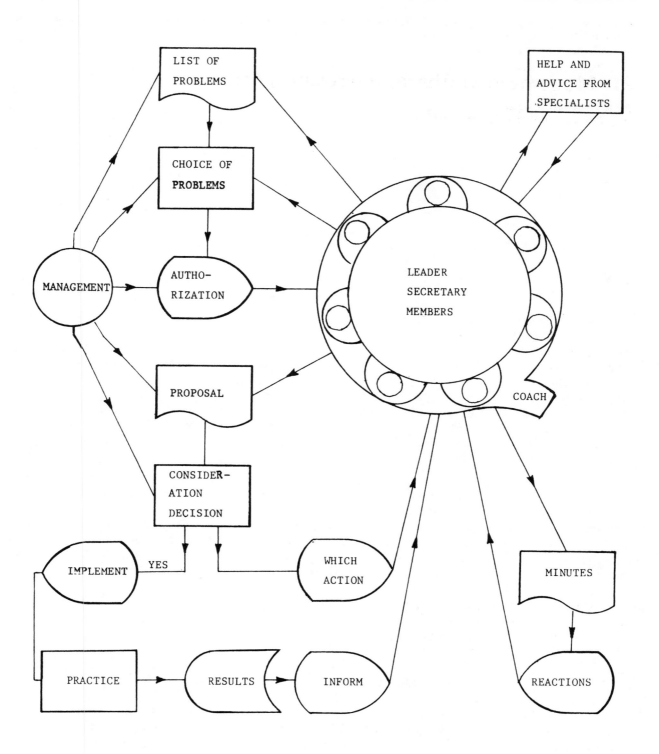

FIGURE 3 -- Interface between QC Circle and Management

ZD Movement in Bharat Electronics Ltd.

by K.M. Shanmugam
B.E., Dip. SQC (ISI), B.E.L., Bangalore

This article describes the experiences gained in initiating the Zero Defects Movement in one of the divisions of Bharat Electronics, Limited. Three years of operation have yielded many good lessons which the author shares here for the benefit of those interested in such Movements. A study is made of the basic philosophy of the ZD Movement, method of initiating, hurdles faced, results achieved and the limitations thereof. Finally, the shortcomings of the Movement are examined and possible improvements suggested in order to enhance its scope and potential value.

QC circles and ZD programme have been successfully functioning as motivational means in Japan and U.S.A. for the last 10-12 years. However, these methods have not spread to different industries in our country. Further, the people who are familiar with these programmes have also, by and large, discarded them as unsuitable for conditions in India. Due to these reasons, there is hardly any major impact of these programmes in this country.

This article highlights some of the salient features of Zero Defects and shows how it was successfully adapted to the local conditions of Bharat Electronics Ltd., for improving quality and productivity.

MODERN INDUSTRIAL SCENE AND ZD

The modern factory is a complex system with separate planning and execution followed by the extreme division of work. While this might have resulted in specialization, it has also created other adverse effects. Workers, for instance, are not able to know the effect of their work on the ultimate product; work, in many instances, has become monotonous, uninteresting and does not utilize or need the full capabilities of men. These have led to a lack of involvement and consequent poor quality and production. Under these circumstances, motivating men for quality or productivity has become a difficult task and is usually approached in two ways.

In the first approach, workers are assumed as a set of lazy, dull and indifferent people who prefer to be led and dislike responsibility. To make them work better, Management has to enforce strict control and direction. Money or financial benefit is considered as the main motivating force under this method.

The second is based on the concept that workers are not by nature indifferent or resistant to organizational needs but have become so now. But they do have the potential for development, the capacity to assume responsibility, and a readiness to change behaviour and these have to be triggered. Management has to recognize and develop these human characteristics by creating opportunities for releasing potential, removing obstacles, encouraging growth and providing guidance.

From the above it is seen that the first approach places excessive reliance on external control of human behaviour, while the second is more positive in nature relying on self direction and control. A change from the first type to the second has become necessary now to get the best out of every individual and in this task the ZD movement plays a vital role.

There is another important aspect for achieving quality or productivity in the modern factory. This is the attitude of mind. It recognized that good quality or high productivity does not emerge merely from good equipments, instruments, gauges or other facilities which are of course important, but from the state of mind of plant personnel from top executive down to the operator. Lacking this state of mind no amount of investment in measuring devices or equipments will yield the desired results. Zero Defect movement provides ample scope in this attempt to develop the right attitude of mind among all.

WHAT IS ZD?

The philosophy of ZD is nothing new and it implies defect free work or doing the job 'right the first time' and this applies to all people in an organization. Under this movement the ZD members are expected to use their full capabilities to improve their own performances and show pride in their work. By creating an awareness of the importance of their work in relation to the final product, their work becomes more meaningful and interesting and results in better performances.

In practice the working of ZD is made more versatile by combining the concepts of 'QC Circles' so that other aspects like improvements of cost, productivity, attendance, safety, etc., are also covered under ZD movement.

ZD works in groups of homogeneous people joined together voluntarily to solve their own departmental problems on quality or productivity. The choice of problems, methods of solution, etc., are left to the members concerned.

ZD works in a system of target-by-target accomplishment. First a target is selected and after achieving it, the next higher target is selected and so on. This makes the movement a continuous working. The achievement of target makes the ZD members self-confident and results in self-satisfaction and brings them recognition. It is these results that fulfill the egoistic needs of men who respond to this movement.

INITIATING ZD

It all began with the visit of a Senior Officer of BEL to Japan, where he was deeply impressed by the tremendous impact of ZD programmes and QC Circles on Japanese quality and productivity. This prompted him to make a detailed study of these programmes apart from his regular assignment. On his return he had detailed discussion with Management of BEL regarding the potential of ZD. Even though there was divided opinion among Managers on the success of ZD, a beginning was made in one of the divisions with the support and encouragement of Management.

This was followed by the formation of an informal organization for the Movement:

> Patron
>   General Manager
> President
>   Dy. General Manager (HPE)
> Vice-President
>   Manager, Fabrication (HPE)
> ZD Co-ordinator
>   Asst. Works Manager
> Managing Committee
> Organizing Committee

The Managing Committee was meant for spelling out the broad policies of working and consisted of Managers of Fabrication, Inspection and Production Control, and the ZD Co-ordinator as members.

The Organizing Committee was meant for steering the actual working of ZD and consisted of all Senior Executives of Shops, Quality Service, Methods, Planning, Work Study and ZD Co-ordinator as members.

The ZD Co-ordinator worked out the complete details of the movement and after discussion with the Managing and Organizing Committee, finalized the course of action. Accordingly the following steps were initiated.

PROMOTIONAL TALKS

Informative posters highlighting the salient principles of the ZD movement were displayed in different places. This attracted the attention of all and raised their curiosity to know more about the movement. At that time a series of talks were arranged in different departments and shops covering about 50-60 people at a time. These talks were normally held in the work spot itself and in some cases were conducted in a lecture hall. The topics covered in these lectures were: philosophy of ZD, importance of the attitude of mind, benefits of ZD, etc. The aspect of self-development and self-satisfaction were emphasized in these talks. The employees were asked to clarify their doubts during these talks so that they could get adequate insight into the movement. No financial or monetary commitments were assured as benefits. However, the indirect benefits like training, recognition, job satisfaction, etc., were impressed upon the people who were curious to know about the benefits of this movement. These talks were extended for about a fortnight covering a cross-section of all the groups of people. At the end of each talk a brief write up of the movement with a 'Pledge Slip' was issued to all. A time limit was set for returning the signed pledge slips. Those who were not convinced about the movement were asked to return the unfilled pledge slips. Membership was by and large voluntary, but a few cases required a lot of education and selling of ZD. Perhaps this is inevitable in any new movement.

ZD GROUPS - FORMATION AND PREPARATION

Pledge Slips revealed that almost all people had volunteered to join the movement showing the enormous efforts put forth in promotional talks. Homogeneous groups of 7-10 members were then formed taking the assistance of respective shop executives. The participation and grouping can be summarized as below:

	Strength Joined	No. of Groups
Production Shops	392	59
Production Control	79	9
Inspection and QC	36	5
Clerical Staff	9	1
Executives	22	2

These ZD groups were then individually addressed by the Organizing Committee members to explain the actual working of ZD. Among other details the two important aspects emphasized in these meetings were the proper selection of ZD group leaders and the targets. These are highlighted here.

<u>ZD Group Leader</u>: Should be a person with immense faith in ZD and with the drive and initiative to lead the members towards achieving the targets. He need not be a superior or an executive but can be even an operator.

<u>Targets</u>: Should be realistic, achievable within 3-6 months, pertain to improvements in quality or productivity, quantitatively measureable and should not involve huge capital outlay or radical changes in existing working conditions. The members were also informed that sincere efforts would be made to provide them reasonable facilities, training and so on to help them pursue their targets. With these guidelines the ZD members selected their group leaders and were preparing to start functioning.

INAUGURATION AND WORKING

To signify Management's interest in and importance of the movement a formal inauguration was held in which the top management and heads of departments participated. The patron (G.M.) of the movement exhorted the workers to make the movement a sustained success. A group leader also spoke and assured the Management their wholehearted participation and efforts.

The ZD groups then selected their targets some of which are exemplified below:

Section/Department	Typical Targets Selected
1. Production Shops/Fabrication	To reduce rejections by 20-30%.
2. Mechanical Assembly/Fabrication	To reduce the assembly returns by 30%.
3. Tool Room/Fabrication	To clear the existing backlog in Milling Centre.
4. Methods/Production Control	To indigenise components/parts at 2 per month.
5. Tool Launching/Prodn. Control	To reduce launching time to 1 month.
6. Plant Engineering/Fabrication	To reduce down time by 15%.
7. Inspection/Quality Service	To reduce inspection errors by 30%.
8. Office Staff/All Depts.	To reduce stationery consumption by 25%.
9. Executives/All Depts.	To increase the efficiency by 5%.

The above targets were planned out on a time-bound programme and were aimed at completion in 3-6 months, when next higher targets would be set. Towards these targets, all ZD members started working. Meanwhile arrangements were made for training classes.

TRAINING

Training classes in Quality Control, Production Control and Work Study, were conducted for different batches of 20-25 members in a phased manner. These classes were of 8-10 hours duration spread over a week during the working hours. The topics covered under Quality Control were: Meaning of quality, responsibility for quality, process control, simple QC methods, quality mindedness and so on. Practical ideas and case studies were discussed to enable the members to approach their targets by systematic collection and analysis of data, and drawing of inferences on the problems.

GROUP DISCUSSION AND SUGGESTIONS

In order to co-ordinate the working and to decide on the ways and means of pursuing their targets, the ZD groups were allowed to meet and discuss for 30 mts. in a discussion room every fortnight. Here the role of group leader was very vital in steering the discussions in the right direction and to get all the members to participate and arrive at the course of action to be taken. In the initial stages, an Organizing Committee member used to guide the discussion. The ZD members were cautioned to utilize this time usefully and solely for the purpose of their target.

Suggestions for prevention or reduction of defects or any other improvements (management controllable) were received as a result of these group discussions. The group leaders formulated these suggestions into a proforma which was considered by the Organizing Committee members for necessary action, the outcome of which was fed back to them through the same proforma. It should be emphasized here that it is of utmost importance to take prompt action on these suggestions to keep up the tempo of the movement, as otherwise the members are likely to lose interest.

EVALUATION OF TARGETS

Evaluation of performances of about 75 ZD groups periodically was an arduous and difficult task faced by the Organizing Committee, since proper records, reports and other data from different departments were required for correct and quick evaluation. Accuracy and promptness of evaluation were both essential to avoid complications and reactions of ZD group members. This called for generation of new records, modifications of existing ones and so on for a number of cases to make the task easier in subsequent evaluations. With these, the performance levels before and after ZD were prepared and notified to all ZD members. For achievements, only non-financial awards like certificates, trophies, badges, etc., were recommended. For failures, the ZD groups either extended the period of time for reaching their targets or fixed a more practicable new target.

It is important that the Organizing Committee members give special attention, guidance and encouragement to the failed ZD groups to succeed in their endeavour and sustain their interest.

GROUP LEADERS MEETING

In a meeting attended by General Manager, and other Managers and Organizing Committee members, the ZD group leaders presented their studies highlighting their work, achievements and views for further improvements. Some of these are summarized here.

A ZD group engaged in assembling precision mechanical drives, gear mechanisms, variometers, etc., was faced with problems of assembly complaints and returns. So the group selected a target to reduce these complaints/returns by 30 percent in three months. The ZD members went through the defects and found them mostly pertaining to sensory quality like 'Movement Hard', 'Operation Sticky', 'Noisy Drive', etc. The members after discussing with the concerned inspectors and assembly supervisors set the required uniform standards for these and adhered to them. Further, they started sending these small drive mechanisms in proper polythene packings thereby avoiding dust collection and consequent friction and improper movements.

Another ZD group from inspection was faced with the task of reducing inspection time for Weidamann press operation jobs involving checking of co-ordinates of a large number (60-100) holes. By retaining the first-off pieces as masters suitably numbered and identified, a bank of masters was built which helped them to reduce inspection time by 25 percent in subsequent batch inspection. These masters were released to production during the last batch.

A ZD group of the clerical staff took up the task of reducing consumable stationery like paper, pencils, carbon sheets, notebooks, etc., and by observing austerity measures, they were able to reduce by 30 percent in first three months and retain the level.

The meeting proved stimulating and was very useful in sorting out the problems faced by ZD groups and also Management to get an overall picture of the working trend of ZD and the areas they could concentrate upon.

RESULTS

The effect of ZD movement on quality levels of different shops has been studied using the quality reports from QC section for two years before and after the introduction of ZD. The Table A shows the details.

TABLE A

RESULTS OF THE ZERO DEFECT MOVEMENT

SHOP	Defective due to	BEFORE ZD				AFTER ZD			
		1970-71 %	1971-72 %	Ave. %	Quality Index *	1972-73 %	1973-74 %	Ave. %	Quality Index *
Machine Shop	Workmanship	2.8	3.1	3.0	48	2.3	3.5	2.9	68
	Total	5.1	7.2	6.2		4.2	4.4	4.4	
Sheet Metal Shop	Workmanship	1.8	1.4	1.6	82	1.5	1.4	1.4	90
	Total	6.7	4.0	5.4		3.5	2.8	3.2	
Mechanical Assembly	Workmanship	1.4	1.2	1.3	74	1.7	2.7	2.2	72
	Total	5.2	5.2	5.2		4.9	5.8	5.4	
Painting and Plating	Workmanship	1.2	0.3	0.8	87	1.0	1.1	1.0	87
	Total	2.6	2.2	2.4		2.8	2.0	2.4	
Rework and Scraps in Hrs. for every lakh of reduction value		66.0	67.6	66.8	--	52.0	75.6	64.2	--

* Quality Index is a measure of standards to actuals of scrap, rework and deviated parts over a period.

It is seen that Machine Shop and Sheet Metal Shop, have shown good improvements, as the average total defectives has come down from 6.2% to 4.4% in Machine Shop and from 5.4% to 3.2% in Sheet Metal. Simarly, the Quality Index (a measure of actual levels of scrap, rework and deviation against std. levels) shows substantial improvements from 48 to 68 in Machine Shop and 82 to 90 for Sheet Metal. However, there appears to be no impact on the other two shops. It is also observed that generally the first year's performance is very good as expected.

A study of the man hours lost due to rework and scrap is also made using data from Work Study. It is noted that during the first year of ZD, this figure had came down to 53 Hrs. per lakh of production compared to an earlier figure of 67.0 Hrs., while in the second year it has not been maintained but increased.

Apart from the immediate quantitative results of ZD, other benefits which should in the long run reflect on overall improvements, are briefly covered here.

1. <u>Self-Inspection and Self-Control</u>

Before ZD, the operators thought inspection department was there to look after quality. ZD created a welcome awareness and made them realize their important role in building product quality.

In fact, consequent upon starting ZD in BEL, a remarkable change came to pass--operators themselves started checking their products and the patrol inspection by QC men was minimized in many areas.

2. <u>Better Team Work and Co-ordination</u>

As ZD essentially worked in groups, better team work and co-ordination were noticed among departments. Departments which were traditionally at war with each other, started coming closer under the banner of ZD. This was one of the important achievements of ZD, as co-ordination of different departments for a common goal is by no means an easy task in a big organization.

3. <u>Better Communication at all Levels</u>

Through ZD, the organizational barriers of communications were bypassed. In the group meetings and discussions the members and the senior executives were freely exchanging their views and this also improved the human relations.

4. <u>Self-Development and Self-Satisfaction</u>

By attending training classes and group discussions, and by striving for achieving their targets, ZD members were able to learn more and increase their confidence to solve problems. Since ZD work was carried out in addition to the normal routine work, there was job enlargement and the members had ample scope to utilize and display their talents. These led to self-satisfaction and self-development perhaps the most important rewards of the ZD movement.

CRITICISM AND OPPOSITION

ZD Movement has been criticized and opposed both inside and outside BEL, on the ground that it can at best lead only to limited results since the programme aims at prevention of operator errors which form only 20 percent of total defects. Firstly, this criticism is not applicable to ZD of BEL where the approach has been to tackle all defects. Secondly, even assuming that ZD has better effects on operator defects, the proportion of operator to pre-production service defects (usually termed as 'Management Controllable') vary from shop to shop in different industries. For instance an analysis of this for the last few years on our Fabrication Shops reveals the following average values:

	Operatior	Pre-Prodn. Service
Machine Shop	57%	43%
Sheet Metal Shop	36%	64%
Finishing Shop	37%	63%
Mechanical Assy.	33%	67%

In areas like Electronic Assembly operations the operator controllable defects are likely to be still higher and hence ZD can prove very effective. Further, the pre-production services like providing good equipments, facilities, instruments, tools, etc., are not lacking in many of the public enterprises. But they seem to lack good team work and co-ordination which may require considerable efforts and time, for a breakthrough. The potential of ZD movement in this direction requires to be stressed.

There was opposition to ZD on other grounds. For some, the concept was not convincing since they believed more in the 'first' approach of motivation by money mentioned earlier. Perhaps these men may lag behind others before long. There were also people who were reluctant to accept the movement on personal grounds and outlook of not following the others. To these, our suggestion is that they can devise their own methods using the basic principles of ZD.

ZD IN OTHER PLANTS

There were a number of enquiries from visitors and dignitaries from different places to study and know about the ZD movement of BEL. One such visit by the Senior Officers of Bharat Heavy Electricals, Thiruchi, had resulted in initiating a similar 'self-development programme' movement in their plant. (This was inaugurated by the BEL President of ZD Movement.) These visits not only helped to spread the movement outside but also gave us a stimulus to continue our efforts further.

CONCLUSION

Considering that ZD movement aims at harnessing the energy, enthusiasm and ingenuity of work force towards better quality and productivity, the movement can well be exploited for our Indian Industries with their vast potential of human resources. ZD will create a good atmosphere for both workers and Management to solve the company's problems on a mutual understanding basis. It is needless to say that in organizing the movement in a plant, it should be planned on a long range basis. For this the following aspects should be considered:

1. Management's acceptance, support and encouragement to ZD is a vital factor for the success of the movement.

2. Before initiating ZD, a lot of preparatory activities to set right the pre-production services to satisfactory levels will prove ZD very effective.

3. In order to make ZD flourish and serve the company on a permanent basis, a proper organizational linking is necessary. For this ZD can be attached to QA or IE department with a senior officer nominated as ZD co-ordinator (and this can be rotated also).

4. It is recognized that if a small proportion of the rework and scrap costs are spent on prevention activities like ZD, much savings could result. Hence Management could think of allocating say 5-10 percent of rework/scrap costs on ZD and watch for results.

5. All supervisors and executives have to be first educated and trained and made to accept ZD. This will make the task easy for further spreading it to operators. In addition, this will also reduce the conflicts if any between the ZD group leaders and the supervisors later.

6. New forms of recognitions, trophies, rewards introduced now and then, will help to keep up the tempo of movement. Failed ZD groups require special attention, guidance and encouragement.

7. The role of ZD is supplementary to QC and IE department in spreading the quality and productivity consciousness among all. In addition, through ZD it would be more easy to extend the activities in areas like planning, R and D, sales and so on where it is otherwise traditionally difficult to enter. Hence, QA and IE departments have to be actively associated with the working of ZD.

8. In big companies, the evaluation of a large number of targets can be a difficult task and in such instances the ZD groups can themselves be allocated this responsibility.

9. As the success of ZD largely depends on the ability of ZD co-ordinator and the ZD group leaders, the choice of the men should be given a careful thought. Faith in ZD, dynamism, practical outlook, sincere and hard work, ability to communicate clearly, etc., form some of the essentials in these men.

10. Top Management by calling for details, attending the group discussions, personally following the implementations, etc., should actively participate in the functioning of ZD. This will stimulate the working and also sustain the interest and success.

Finally, whether ZD Movement will survive and flourish in years to come or not, the vision it has kindled so far in BEL and elsewhere continues to suggest that progress in quality and productivity can spiral by this movement, if a fair trial is given.

(The views and opinions expressed here are those of the author and not the organization.)

Some Trends in Productivity and Quality of Life

by W. Maurice Kaushagen

In management of technology in the late 20th century period, many trends are identifiable. With the importance of technology in today's world, these industrial/technological/sociological trends present great challenges and opportunities to management. For example, a proper balance between productivity and quality of life is an elusive goal. The writer had a unique opportunity to study some of these trends and developments in the automotive production line setting in northern European countries: Sweden, Germany, England, and Norway. The following report is a summary of these studies. Clearly, traditional production line problems involving boredom and labor unrest can be ameliorated by job enrichment and teamwork organization which modifies cycle times* from about one minute to as much as 30 to 50 minutes. These trends and their impact on engineering management are worthy of careful study.

INTRODUCTION

An important area of management development, particularly in technological fields, is that of striking a proper balance between productivity and quality of life. It has become more and more evident that workers in the highly technical production line setting often feel unhappy about their lot in life. For some years now we have been observing a growing trend towards job enrichment, job enlargement, job satisfaction, and related phraseology ultimately aimed at balancing productivity with quality of life. One can consider these phenomena in a quasimathematical form as follows:

$$\text{JOB PERFORMANCE} = \begin{Bmatrix} \text{ABILITY} \\ \text{(knowledge)} \\ \text{(skills)} \end{Bmatrix} \times \begin{Bmatrix} \text{MOTIVATION} \\ \text{(interest)} \\ \text{(need satisfaction)} \\ \text{(social environment)} \\ \text{(physical environment)} \end{Bmatrix}$$

- o Output
- o Low Turnover of Personnel
- o Low Absenteeism

- o Natural Ability
- o Education
- o Training

- o Pay
- o Fringe Benefits
- o Job Interest
- o Security
- o Perceptions

---

* Cycle time is the time interval during which each production line worker must do his specific repetitive task.

The current efforts build on the work of Douglas McGregor and others who speak of physiological needs, social needs, and egotistic needs of people.

When one considers productivity, motivation, job interest, and worker morale, it is important that the objectives of the parties involved be carefully identified and understood. Objectives can be identified in terms of various entities: the individual, the corporation, the union (representing the individual (?)), the nation, and society at large. Clearly, there is compromise required because all objectives cannot be achieved either in the short term or the long term.

There are interesting and important trends with respect to setting objectives and in the achievement thereof. In some segments of society there is characteristically a strong adversary approach such as in the stiff bargaining position between union and management. This frequently leads to an impasse with strong emotional and sociological conflicts leading to strikes. The interesting variation observed by the writer in the recent series of visits is most evident in Sweden. The trend here is to a reduction in the adversary situation, i.e., a more cooperative spirit. Thus the parties involved are able to explore jointly a more balanced alignment of objectives. This leads to the identification of a much broader field of "quality of life" elements than the obvious salary increase or more holidays or shorter work weeks. Thus we have emerging in Sweden a much more careful study of the worker environment, the elements of his job that lead to boredom and frustration, and the worker participation in identifying objectives and ways of achieving them.

During the summer of 1976, the writer had the unique opportunity of visiting several companies in Norway, Sweden, Germany, and England. Managers were pleased to discuss these topics and to demonstrate results with actual on-line detailed plant tours. This series of visits was extremely interesting because of the different settings in which these plants operated: national, sociological, and geographical. Thus the writer gained a wealth of insight and understanding about the way in which modern trends in management of technology are being implemented in the auto industry.

A BODY PLANT

This plant had been producing automobiles with the "traditional" production line approach for some years. Problems had been developing in the late 60's however. The manifestation of these problems was in labor shortage (unemployment is very low in Sweden), and a need therefore to import production line labor from Finland and other countries. As production line objectives became more intense, absenteeism and employee turnover rose sharply to the point where annual turnover figures began to approach 50 percent. Clearly this was an impossible situation and the company began to experiment with pioneering production line management concepts such as teamwork organization involving 20 to 50 minute cycle times, group self organization, new plant layouts, and much broader responsibilities laid on the individual worker. A spectacular demonstration of the conviction as to the correctness of this approach occurred in July 1975 when the body assembly division was shut down for the annual July vacation, and the assembly line was completely renovated in preparation for a new production line in August. During that one month of effort the old three minute cycle time in the body production line was modified to 50 minutes, and parallel production loops were set up for teams of four to 20 people. These teams have a very extensive list of responsibilities: control of incoming materials, control of sequence of operations within the team, training, permission for one day leave, meeting production quotas, direct control over jig servicing, substantial responsibility for inspection, which the obligation of redoing the work if later inspection yields a reject. The emphasis here is on a "little autonomous factory" for each group.

The manager responsible for this effort was very enthusiastic about the substantial reduction in turnover. Absenteeism was reduced (although this is still a problem), and the very substantial reduction in sensitivity of the production line to individual absenteeism was made. It is quite possible with the self organization, for example, for a group to be missing one person out of five, (occasionally), and still be quite effective in meeting monthly production quotas.

The writer was given a wealth of literature including several books published by the Swedish Employer's Conference and by a joint workers and managers organizational group studying these problems together. It is significant that joint publications are forthcoming in this manner.

ENGINE PLANT (FIRST)

At this engine plant the four cylinder, two liter engine is being produced under an organizational arrangement of assembly loops with four or five members in the team and 30 minute cycle times. About 110,000 are produced per year or 440 per day. This organizational arrangement has clearly improved worker morale and has brought down the high annual turnover rates to manageable levels. Again, Finnish workers have been imported to satisfy the labor needs. Again, the engine plant was organized for teamwork efforts within an older production plant which was substantially modified. It is interesting to see that the pace was very relaxed. In fact, it was evident that the self organization of the groups made it possible for one or several members to take a coffee break in an adjacent plant lounge and in no way hinder production. The contrast with the modern short cycle time engine plant in the U.S. is very impressive. Further, a large number of women were employed and fully intermixed with men. Because of highly developed handling equipment, lifting of heavy pieces is completely eliminated and men and women are interchangeable as far as the tasks are concerned. Each team member is capable of doing any of the tasks of the team. Thus flexibility of assignment and internal self organization is the order of the day.

The management of both body and engine plants were enthusiastic about their progress to data and were now in the process of establishing the most quantitative evaluation that they can under the varying circumstances. Clearly, worker motivation is a function of the penalties associated with not working. It turns out that these penalties are surprisingly small in Sweden's highly socialized benefit program. Thus, one can say that meaningful data on productivity is not easy to obtain, because of the impact of the changing social environment.

ENGINE PLANT (SECOND)

The plant was built from its inception with team efforts in mind. The entire factory has a layout of four production wings extending out from the long assembly and test building. Between each of the wings is a large garden area. The whole plant is light and airy with windows and relaxation areas in profusion. Each engine is set-up on a small self-propelled vehicle that is controlled by cables embedded in the floor which are driven by a master computer. The assembly is done in a small workshop atmosphere with buffer space so that production flexibility in the team loops can be arranged. Cycle times of about 20 minutes are typical. Thus the whole layout is very group oriented and the atmosphere is one of environmental well being, both physical and psychological. There is a very large amount of automated equipment for machining and handling as well as for measurement.

KALMAR PLANT

This is perhaps the most famous of the Swedish plants with the new philosophy of production lines. To quote from Pehr G. Gullenhammar, Managing Director of the Volvo Group: "It takes patience, time, and effort to change a working organization, to modify working tasks, and to make up new work planning. I believe that the human factor and efficiency can be combined. I also believe that in the community of today these two factors are inseparable. At Kalmar we have endeavoured to arrange car production in a manner which will make it easier for the employees to find meaning and satisfaction in their work. This is a factory which, without sacrificing efficiency and economic results, enables the employees to work in groups, to communicate freely, to carry out job rotation, to vary their rate of work, feel identification with the products, to be aware of quality responsibility and also to be in a position to influence their working environment. When a product is made by people who find meaning in their work, it must surely be a product of high quality."*
This statement sets the stage for the bright and shiny Kalmar facilities.

In the Kalmar plant with its basic hexagon geometrical motif, the automobiles are set on self-propelled platforms controlled from a central computer driving cables laid in the floor of the two level plant. Assembly teams work in a relaxed atmosphere with about 20 minute cycle times and coffee breaks in the adjacent relaxation lounges can be organized into the cycle time by the team members. The plant is very quiet. The large assembly platforms move quietly and smoothly. It is indeed an impressive setting for the manufacture of automobiles.

GERMANY

In Germany, the writer visited a very large, multi-model automotive production facility. Employment approaches 50,000 people and it was indeed an overwhelming place to visit. The writer was most graciously received and given an extensive tour, lengthy discussion, and lots of literature. It was interesting to observe the relationship between the sociological traditions of Germany as compared with that of Scandinavia. Productivity was obviously very high and seemingly worker morale was very good, although modern production line management techniques as covered above were not as much in evidence. As one studied the production activities however, there was substantial evidence of a form of group activity with corresponding flexibility and work rotation. The overwhelming impression, however, was one of traditional German mechanical industriousness. The assembly setting certainly was also one of a

---
* VOLVO Kalmarverken, The Volvo Kalmar Plant, AB Volvo, Kalmarverken, Informationsavdelningen

production line with group activities along the line rather than the parallel loops of Sweden. Germany does not seem to be faced with the same worker shortage that was evident in Scandinavia. In fact, the labor pool situation tends more towards a level of unemployment more typical of the U.S. Thus there is a sociological motivation that is more nearly characterized by job shortages than by worker shortage. In summary, it can be said that this plant seemed to be fully aware of "quality of life" matters but was dealing with it "in their own way."

ENGLAND

Quite a different situation exists in England. In 1975 there was an extensive investigation by a governmental committee headed by Lord Ryder. The report coming from the committee addressed the problem of solvency and the very existence of private enterprise in the automotive companies involved. Thus the impression was gained that such topics as job enrichment leading to increased productivity had to be set aside in favor of basic survival (in a free enterprise sense).

It is interesting to observe, however, that there are said to be fewer strikes in the U.K. They are certainly in a job shortage mode and perhaps that is the reason. There are other differences, however. The local shop steward is a very prime force in union negotiation with management. Thus contracts are negotiated on shop by shop basis rather than broadly by United Auto Workers as in the U.S. This is both a blessing and a curse.

For example, while the writer was in England, a new model (said to be revolutionary) was scheduled for release to the public. As they were moving to the release date with a first production run of automobiles, the assembly plant struck. That strike was settled by some unique features including a lottery of some of these automobiles to the employees. When this settlement became known to supplier plants to the assembly plant, they asked the question, "What's in this for me?" and as the writer left England a wave of related plant strike situations was unfolding. Clearly, the British have industrial problems of a fundamental sort.

SOME PERSPECTIVE

Clearly the matter of motivation, job enrichment, and productivity is very complex. If one thinks about a set of relationships in terms of feedback, some additional insight is gained. Figure 1 describes, however simplified, some of the relationships. For example, compensation feedback for employment is compared in the person's mind with "compensation for unemployment" as indicated by the two-position switch at point A. In Sweden it seemed that the "compensation for unemployment" plus the basic worker shortage were very important factors that motivated automotive production managers to increase job satisfaction. The quantitative analysis of cause and effect is being studied extensively in Sweden but it is very evident that in a complex multifactor situation like this is, accurate measurements are very difficult.

The relationship among some of these factors can be looked at pictorially in an oversimplified three dimensional diagram, Figure 2. Here we see, for example, that point S is indicative of the situation in Sweden where job enrichment is being emphasized, where penalties for unemployment are not so very

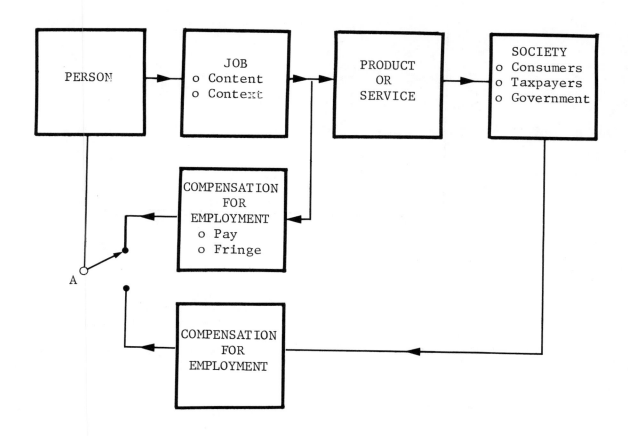

FIGURE 1.   WORKER SATISFACTION FEEDBACK

Factors: o Job content (work itself, interest, challenge....)
           o Job context (environment, pay security....)
           o Job shortage or worker shortage
           o Penalty for unemployment

high and where there is no labor surplus. Point G shows a situation more typical of Germany or the U.S. where the penalties for unemployment seem higher, where job satisfaction is emphasized less and where there is a large labor surplus.

Both Figures 1 and 2 are oversimplified. They are presented, however, so that there can be more clear visualization of the interrelation of these matters. A further extension would include consideration of the overall cost of production in the several companies involved. We, in the U.S., can get some feel for this dimension by noting the relative price (current U.S. dollars) for automobiles produced in the several companies under consideration.

It is evident that important trends in production line management are underway in many quarters. How these will unfold in the future is of great

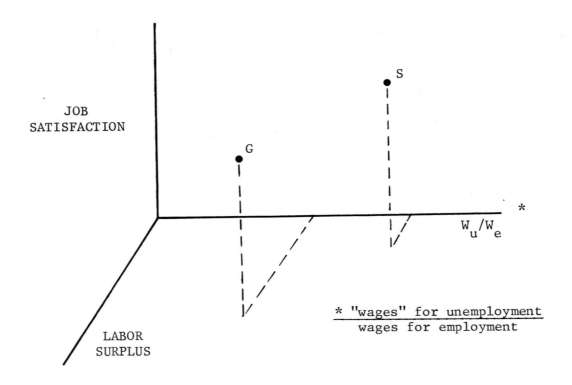

FIGURE 2. THREE DIMENSIONAL VISUALIZATION

interest to managers responsible for effective production in a solvent environment.

The author is with the Technology Division, State University of New York, Buffalo, NY 14222.

Applications of Psychology to Quality Control*
by Dr. Paul Peach, PH.D.

In the course of his activity, the quality control specialist must become a jack of many trades. If he works in a textile mill, he must learn that the word "silver" has a special pronunciation and meaning different from the one he is used to. If he works in the field of chemistry he can discover that there is a substance called periodic acid, pronounced purr-I-owe-Dick. In a metal shop he must learn about jigs and fixtures, and about the properties of various metals and tools. Though he will not likely become a professional in any one of these fields, yet he will have to learn enough about it and its vocabulary to talk to the professionals in their own language and understand their problems. The QC man, and in fact the statistician, in general, never finishes learning.

However, relatively few statisticians have ever been exposed to the discipline of psychology. One might wonder at this. Psychology is that branch of science that studies human behaviour, and since the QC man, like all other workers, must deal with human beings, one might suppose that a knowledge of at least the elements of psychology would be part of his necessary equipment. I think the reason for this not being usually accepted is, that psychology is an inexact science, and moreover one in which it is difficult to separate actual, objective observation from subjective evaluation and judgment. If not exactly true, it is at least partially true that two psychologists, using the same data, can arrive at diametrically opposite conclusions. Statisticians not working directly with psychology thus tend to suppose that anything so inexact can add little of value to their discipline.

Now, it is certainly true that psychology is an inexact science; but so are physics and astronomy. The degree of inexactness is of course not the same for all these; in physics we may announce a conclusion with (say) 99.99% confidence, whereas in psychology our confidence may be less than 1%. The difference is one of degree, but that is all; there is no such thing as an "exact" science, and because of the Heisenberg Principle we will know that there never will be.

---

* Based on a talk given under the joint auspices of the IAQR & SQC Unit, ISI, Bangalore. Till recently Dr. Peach was Ag. Director, Computer Centre, University of Malaya, Kuala, Lumpur.

INEQUALITY AMONG MEN

The main defect of psychology is that it is too easily coloured by the facts that (a) it deals with human behaviour, and (b) we, psychologists included, are ourselves human beings. They tell the story of the psychologist who wished to study the behaviour of a chimpanzee. He accordingly led the ape to a room, closed the door, and immediately applied his eye to the keyhole. All he could see was another eye: the monkey had the same idea himself. We can never know how we might look to a super-intelligent monkey. Now, we have all grown up--more or less--in the democratic tradition: All men are created equal.

But it takes no profound analysis to show that though we may agree that all men are endowed (in Jefferson's phrase) with "certain unalienable rights; that among these rights are life, liberty, and the pursuit of happiness; that to ensure these rights, governments are instituted among men", yet men are not in fact created equal. I am not the intellectual equal of Aristotle; I cannot paint pictures like those of Michaelangelo and Leonardo; I cannot run a mile in 4 minutes; in my view, and borrowing from the language of computers, every human being is born with a program already resident in his mind. As he matures, additional programs will be written and stored in his control area: these programs may, and in fact, doubtless, will eventually be the main determinants of his actions. Yet (again in my view) the original program is never erased; it may well be that in ordinary circumstances it will never be called into operation, but if by some drastic change in the environment the acquired programs become ineffective, the inborn program will take over.

If we accept this point of view, it becomes of great practical importance to learn about these inborn programs. If we admit their existence (and many sociologists do not) it seems inevitable that they are genetically determined. It becomes important to ask: Are they the same for all people? (To this question I think we can rather confidently answer No.) Are there certain elements that are common to a great many people, or perhaps all? I should answer Yes. If I stick a pin into a baby's bottom, it (the baby) will make a sound, and this sound will be about the same whether the baby is Chinese or American, or whether its parents are Communists or Prohibitionists. The baby could not have learned this response from its environment; it is part of his genetic programming. It is a proper domain of psychology to study human beings and discover, if possible, whether these programs exist, and which elements of these programs are widely or perhaps universally distributed. And to the manager, whether in Quality Control or any other field, it is important to know these qualities, because he himself is human, and he is dealing with human beings.

PROBLEM OF MOTIVATION

A main problem in all management is that of motivation. Some assembly line operations are simple enough so that a chimpanzee can be trained to do them; I believe the experiment has actually been tried. By offering the monkey a food reward, he might be induced to stay on the job a little longer; but there is a limit to the number of bananas any ape can hold, and once he is full, the promise of an additional banana will not motivate him further. In nearly all aspects of management, motivation is an important problem. It would be a help if we could find some one motivating factor capable of influencing all people--a portion of the genetic program universally, or almost universally, present. In the past it has been assumed that the universal motivating factor is money.

I wish to argue that money, far from being a universal motivating factor, is a very poor choice in the majority of situations. We all know that there are people who have more money than they can count, and we know that most of them are greedy enough for more money. But (leaving true misers aside) are these people really seeking more money as money? I think not.

THE PEER GROUP

My suggestion is that the motivating factors capable of influencing at least a majority of human beings are (a) the desire for one's own self-esteem, and (b) the desire for the esteem of others: in particular, of what sociologists call one's peer-group. Your peer-group consists of the cultural group or subgroup to which you belong. In U.S.A., young people of High School age (say 15 to 19) form a very close-knit peer-group. These young people set behaviour standards for the group. They decide what kind of clothing will be acceptable, what kind of music they will listen to. It is of course perfectly possible for a young person to disregard the laws of the peer-group, and he may not even suffer much of a penalty. He may prefer to shave rather than grow a beard; to read books rather than go to teen-age parties. He will not exactly suffer ostracism, but in a sense he will ostracize himself. He will not be subjected to insult--in fact, he may be respected and admired--but he will simply not be a member of the group. Many young people in America accept this kind of separation with indifference, but the great majority want to be members of the group. For them, acceptance by the group is more important than other alternatives.

We come now to the factory worker. His peer-group will usually be his fellow-employees. In a larger sense, he may identify with the working class, as opposed to the "bosses" or "capitalists". Still more largely, he may identify with the culture in which he has grown up. If we try to influence him in directions approved by his peer-group, or by his tradition, we shall usually find our task quite easy. The mere promise of winning the approval of his peer-group will usually be enough to influence him. If, however, we attempt to influence him in other directions--even in directions not actively in conflict with his group--we have to offer him incentives. The incentives most likely to work are those that make use of his inborn behaviour programs: that appeal to his self esteem, and to his desire for approval.

At the lowest level, an incentive can be provided by money. If people are very poor (I know about this, because I came from a background of poverty) a little extra money is a powerful incentive. At this level, people regard money not only as a means of getting enough to eat, but also as an index of social status. But as we rise in the scale, money ceases to be a measure of status. The level may be different for different people, but at some level, for every person, money for its own sake ceases to be an important motivating factor, and status--the esteem for one's own self, and the respect of one's peer-group--takes its place.

Now, American industry is largely organized around mass production. In most mass production operations, the workman is performing a tedious job. He probably knows or suspects that a monkey could do it as well, if the monkey could be motivated to do it eight hours a day. He may have little or no comprehension of how his task fits in with the production of a finished product. What does such a task offer him in the way of satisfaction? It may offer him money; but if he is a typical American worker, in ordinary times this may offer him very little indeed in the way of satisfaction. I suspect that at least many of our strikes arise, not because of protests against low wages or bad working conditions, but because of the need of the men to flex their muscles: to prove to themselves and to each other that they are not merely faceless units, mere interchangeable cogs in the production machine, but human beings, individuals, men and women with personalities and special business and desires--free and independent members of the community who, if they are not allowed to express their individualities in a creative way, will express them in whatever way offers. "I am a person, unique, important," they are trying to say. "If you don't believe it, see how easily I can shut down your factory or for that matter your whole economic structure."

QUALITY MOTIVATION

How, then, can we motivate human beings to do what we want them to do, and in particular to work to quality standards? I think we must start by admitting that monetary rewards in themselves will not do the job; we must somehow appeal to the workman's desire for self-approval. In some cultures--the German and Scandinavian, for example--quality workmanship is traditional; it takes little or no inducing to get workmen to maintain quality standards, because for any workman to get a reputation for careless or inferior workmanship would quickly earn him the dis-esteem of the other members of his culture. When there is no tradition of quality workmanship, management has a problem. How to induce a workman who, to be blunt, does not care a tuppence whether what he produces is good or bad--how to induce him to produce the kind of quality you want?

My suggestions can be inferred from what I have said already. We must appeal to (a) the worker's self-respect, and (or) (b) to his desire for the approval of his group. Just how this is to be done must be decided for each establishment. We might ask: How does the worker risk losing the esteem of his group? One way is by causing them extra work or trouble. If by defective workmanship the operator makes the next man's task more difficult, it may be possible to make him conscious of this by placing him temporarily in the next man's shoes. Thus, in a sequence of operations where A feeds B, B feeds C, and so on to E, there may be some profit in letting a worker who will eventually do Operation A work for a time at Operation E, especially if this can be done with a careless operator at Operation D. The worker, having to cope with the output of this careless colleague, may acquire an understanding of how his own carelessness may make the lot of his friends more difficult. The Volvo factory in Sweden has tried the experiment of giving large segments of the assembly task to teams of workers, who allocate the separate tasks among themselves as they please and vary assignments also; this system seems to work for Volvo (a quite high-priced car) but Detroit has come (rightly or wrongly) to the conclusion that it will not work in Detroit. (We cannot take this as proved; most people automatically resist change.) Various factories have tried various kinds of propoganda methods to make the worker more quality conscious; such methods are often mechanized (for example, a film showing the relationship of parts to a finished product) and since the worker's participation is often merely passive, it may have little effect. The main principle here is that when a person takes an active part in some enterprice, he finds it fairly easy to identify with it; if he is no more than a spectator, he may get bored with watching, or even take the position that the whole exercise is mere propoganda (as indeed it probably is).

The essential point is, I think, that we must recognize that if a job bores the worker, he will take no pride in it. Somehow, we must manage to keep his interest alive. If this means having him do several different jobs in the course of a day rather than staying at one operation for 8 hours, it is still worth exploring. If the worker can see his contribution to the final product as something important and essential, we can perhaps appeal to his self-respect. We cannot do this by verbal persuasion alone; aside from the message content, any message from "the bosses" to the workers is automatically suspect. It may be as well to recognize the fact that if few workers have quality as a goal, few "bosses" care much about the welfare or happiness of their workers and the workers know it.

PSYCHOLOGICAL TESTS--VALIDITY

Aside from the factor of motivation, we have something to learn from that part of psychology which is concerned with tests and measurements. We have, for example, generalized intelligence tests, supposed to measure a subjects I.Q. We have also aptitude tests, intended to predict whether a worker will succeed at some task we may want him to perform. In particular, in an aptitude test, we may ask the candidate to perform certain operations which have, on the surface, nothing to do with the intended assignment. The psychologists have accumulated many files of data on tests of this kind, and since an aptitude test is analogous to raw materials inspection, we may be able to learn something from it.

In an aptitude test the subject is asked to perform tasks--for example, sort coloured beads into separate boxes, pick up small objects and manipulate them in some way, take a simple assembly apart and then put it together again, and so on. For each subject, a set of scores is kept. Later, when the subject is assigned to a regular job, his performance may be scored. The ability of the aptitude tests to predict performance is called the validity of the test measured by a correlation coefficient, represented by the symbol r. If a set of subjects is given an aptitude test, and later their job performance is evaluated, the correlation coefficient (here called the validity) measures the ability of the test to predict performance.

In industrial testing we have an analogue in the use of non-destructive methods in connection with destructive testing to predict the performance of the article tested. Unfortunately, the validity of most such tests is low. In psychology, a validity of .20 is regarded as very good--or perhaps I should say, as about the best we can expect.

Now, one trouble with the statistic is that it is not lineraly related to the variations measured. The statistic of interest is $1 - r^2$, which measures the amount of variation not predicted by the test. If $r = .2$, $1 - r^2 = .96$, and this means that the test leaves 96% of the variation unaccounted for. If performance prediction in industrial testing is to prove its merit, it must clearly do better than this.

One way in which we might seek to improve the validity of a test would be to devise two different tests and combine them in some sort of prediction equation. This approach in psychology is not very successful, because it is

difficult to design two tests that do not overlap to a considerable extent--
in other words, a pair of such tests does not give independent results, and
so the scores cannot legitimately be added. Graphically, suppose we have a
rectangle whose area is 1, and suppose that the amount of information from a
test is indicated by a figure (say a circle) inside the rectangle, and that
the area of this figure is r^2. The points included inside the circle represent the elements of variation that the test can account for. If we add a
second circle, based on information from another test, and if the two circles
do not overlap, their information is additive; but if they intersect and thus
have a portion of their area in common, the information provided by the two
tests is less than the sum of the information amounts provided by the same
tests singly; we are counting some of the information twice.

Because the human mind is subject to many causes of variation, and since
we know very little about how these cases are linked together, it is very difficult to design two tests that do not have a considerable overlap. In the
field of industrial testing we are in a somewhat better position, and statistical analysis does in fact enable us to estimate the extent to which two
tests provide duplicate information. With patience and persistence, aided
by imagination--the most important quality a scientific man can have--it
should be possible in any actual case to increase the battery of bench tests,
eliminate the ones that merely duplicate other tests, and so in time work toward a situation in which non-destructive testing can tell us enough about a
product to enable us to predict its performance with a fair degree of certainty.

An important aspect of psychological testing is that the test may itself
cause a change in the item tested. Every teacher knows that certain students
who do well in the recitation hall or laboratory may make quite low marks on
an examination. The student may become frightened and nervous, and so the
test may quite fail in measuring his scholastic achievement. There is an
analogy in industrial testing. A test may, for example, consist of a mechanical stress or a current overload. It is perfectly possible that the component may pass the test, but at the same time become weakened to such an extent that on the final performance test the component fails. The QC personnel need to be aware that even when a test is, on the surface, non-destructive, we cannot assume that it has no effect upon the product without definite evidence.

CONDITIONED RESPONSE

A final example--for we could continue drawing analogies from psychology
for many more pages. One of the common tools in psychological testing is the
questionnaire. Here the subject is asked to answer a battery of questions,
and some kind of conclusion is sought to be drawn from his answers. Now, consider the following series of questions:

1. Did your mother come to your house yesterday?

2. Did your friend Mohan come to your house yesterday?

3. Did anyone else come to your house yesterday?

By way of comparison, consider the following questions:

1a. Did an electrician come to your house yesterday?

2a. Did the dustman come to your house yesterday.

3a. Did anyone else come to your house yesterday?

Questions 3 and 3a are identical, but very often respondents will give different answers to the two. Whenever we use a questionnaire to gather information (and this will be true, for example, when we seek to measure consumer satisfaction by interrogating users of our product) each sequence of questions creates a context (the psychologists call it a <u>set</u>, meaning a frame of mind) and the question is answered in a way conditioned by this <u>set</u>. I knew of a case in Kuala Lumpur where, by mistake, the same question appeared twice, in different positions in the questionnaire; about half the respondents gave different answers--perhaps because the question itself was ambiguous, perhaps because by the time they encountered the question a second time their phychological set had changed and they were answering from a different point of view.

If is not possible, of course, for the statistician to take all human knowledge for his province. But the QC manager, and indeed the manager generally, deals with people and he does well to learn some thing of what is known about human behaviour. This process of learning should never end.

CHAPTER 7

QUALITY ASSURANCE AND MANUFACTURING PRODUCTIVITY

Quality And Productivity—Interactions

H.J. Bajaria
Lawrence Institute of Technology
Southfield, Michigan

Keywords: Productivity, Quality, Management, Manufacturing Productivity, Communication

Quality - a key to productivity may sound like an old idea, but it has not taken a grassroot position yet in the productivity improvement process. As a result, a production planning process treats the subject of quality very loosely and at times neglects it altogether. Also, the process fails to realize the interactions between quality and productivity, when it tries to improve manufacturing productivity.

The author examines the interactions between quality and productivity for different manufacturing situations, for forming a strategy for productivity index improvement. Also, the author identifies the productivity, business and quality of worklife indexes that are directly or indirectly affected by quality function and that form the communication link between management and quality function.

1. INTRODUCTION

In general, productivity may be defined as an amount of usable commodity, whether it be profits or number of usable goods, per unit of effort spent. Unit of effort can be measured in number of people, amount of dollars spent, square feet of floor space occupied, etc. depending on a particular situation at hand. There are different indexes for different industries at several different levels that measure productivity. Each manager looking at these indexes selects those indexes which can be influenced by his or her area of responsibility. Each manager then specifies in his MBOs to improve these productivity indexes over the specified time span. Once these objectives are achieved he then performs those activities that will help sustain the indexes at the achieved levels.

Top management, however, looks at overall picture and deals with several more indexes than the individual manager. Because appropriate resources need to be committed in raising or in sustaining any individual index.

Quality function has twofold purpose in improving or in maintaining any productivity index, namely, (1) directly influence the index and (2) indirectly influence the index. This type of approach then becomes easy to communicate to top management, because it defines the role of quality in business. What are these indexes that quality function can directly or indirectly influence?

2. RELATION OF QUALITY WITH PRODUCTIVITY IMPROVEMENT

Before examining which of the productivity indexes are affected by quality function, let's examine the prevailing understanding of the relationship between quality and productivity.

When management talks about productivity improvement, they generally refer to rise in "quantity produced per unit of effort." This definition can be made explicit in the sense that it should really say "usable quantity in customers' hands per unit of effort." Let us take an hypothetical example to show what can happen in any productivity improvement situation. Figure 1 examines the relationship between quality level and productivity index for the given quantity of goods produced by the existing production process.

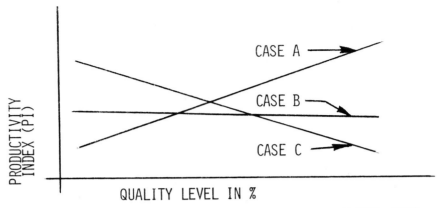

FIGURE 1 - PRODUCTIVITY INDEX VERSUS QUALITY LEVEL

Case A indicates that if quality is improved for the given quantity level, productivity index will improve. This also suggests that the expenditure for a quality function will be justified.

Case B indicates that if quality is improved for the given quantity level, productivity index will remain at constant level. This is the lowest extreme to which the expenditure for quality effort will be justified.

Case C indicates that if quality is improved for the given quantity level, productivity index will go down. This suggests that the quality efforts expended will be in much higher proportion than the productivity raise they will produce.

There are several reasons for cases B and C to exist. Among them are:
(1) Quality system planning is either absent or poor.
(2) People involved are less than competent.
(3) Product design is not consistent with state-of-art.
(4) There is more emphasis on "detecting bad stuff" rather than "preventing bad stuff."
(5) Management is using quality department as a "window dressing" effort; i.e. motivation to improve quality is absent.
(6) Repair and rework are more expensive than the original production piece cost due to inherent nature of the product.

Figures 2, 3 and 4 depict the conventional approach to productivity improvement versus quality approach to productivity improvement for three different situations.

Figure 2 indicates that productivity index can be raised from PI_1 to PI_2 by two alternates: (1) Conventional alternate A - Install a new process and maintain the same quality level; (2) Another alternate B - Increase quality level from A to B.

Sometimes, production management thinks that with the installation of a new process, an operating point will be ②' and not ②. This is, in large part, an erroneous assumption. Higher quality level can be inherent to a new process but it needs to be maintained at that level by quality system planning and procedures carried out by competent people. So, even if the operating point was ②' initially, it will slide back to some lower level ② without adequate quality efforts.

Figure 3 indicates that a productivity index can be raised from PI_1 to PI_2 by only one alternate and that is to install a new process. The cost of any quality improvement efforts will far exceed the productivity raise that can be realized.

Figure 4 indicates strong interactions between quality, productivity and a traditional method to improve productivity. With existing process, it is possible to gain productivity improvement by raising quality level from A to B. However, if the new process is installed, it shows that the productivity goes down with the improvement in quality level. This interaction must be properly understood before selecting alternates for productivity improvement. For example, Alternate 1 - one can go from quality level A to quality level B by spending dollars in quality planning and im-

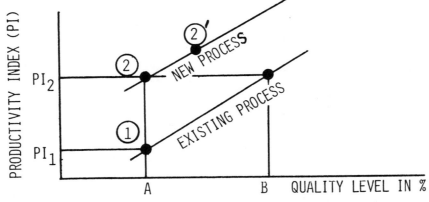

FIGURE 2 - PRODUCTIVITY INDEX VERSUS QUALITY LEVEL
COMPARISON BETWEEN EXISTING AND NEW PROCESSES

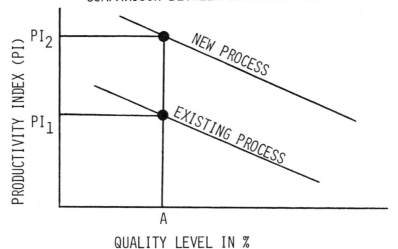

FIGURE 3 - PRODUCTIVITY INDEX VERSUS QUALITY LEVEL
COMPARISON BETWEEN EXISTING AND NEW PROCESSES

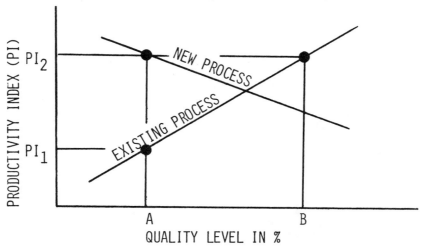

FIGURE 4 - PRODUCTIVITY INDEX VERSUS QUALITY LEVEL
COMPARISON BETWEEN EXISTING AND NEW PROCESSES

provement for the existing process, or Alternate 2 - one can go from from an existing process to a new process and maintain quality level at A.

It is important for a quality function to determine, which of the three situations described above is prevalent before committing resources toward productivity improvement.

3. PRODUCTIVITY INDEXES THAT CAN BE INFLUENCED BY A QUALITY FUNCTION

Now let us return and discuss those indexes that quality efforts can directly or indirectly influence.

Some of the productivity indexes that are easily measurable and can be directly affected by quality function are as follows:
(1) Scrap costs/total sales
(2) Rework costs/total sales
(3) Warranty costs/total sales
(4) Product liability costs/total sales
(5) Lost production costs due to unscheduled downtime/total sales
(6) Quality department costs/total sales
(7) Number of design changes/total sales
(8) Testing costs/total sales
(9) Other indexes

Some of the productivity and business indexes that are easily measurable and can be directly influenced by quality efforts are as follows:
(1) Percent market share
(2) Profit/employee
(3) Profit/Share
(4) Total sales/employee
(5) Number of production hours/unit of product
(6) Maintenance cost/unit of product
(7) Other indexes

Some of the quality of worklife indexes that are measurable and can be indirectly affected by quality function are:
(1) Hours spent in prevention activities/total hours
(2) Proportion of absenteeism
(3) Number of employees terminating jobs/year
(4) Other indexes

4. SUMMARY

When top management begins to understand the role of quality function in productivity improvement and the interactions that can exist among the competing alternates, only then it will begin to desire and support quality efforts.

For a quality function to play a true managerial role, it must communicate its efforts with respect to how it can affect productivity, business and quality of worklife indexes. The author identifies the traditional and nontraditional indexes that must be tracked in order to measure the effectiveness of quality efforts. Of course, one can perceive many other indexes depending on the nature of the business.

Quality efforts can be sold on two basis: (1) Those that are necessary to raise the productivity indexes and (2) those that are necessary to maintain the productivity indexes.

What is basic to quality efforts is to determine the proportions of prevention, assessment and failure efforts that are expended in achieving the productivity improvement task by the quality function. These proportions are unique to a given situation and require an indepth analysis for their determination. Such an analysis is of considerable value in achieving and maintaining the level of productivity indexes.

5. BIOGRAPHY

Dr. Bajaria has been an Associate Professor of Mechanical Engineering at Lawrence Institute of Technology since March 1978. He is also the principal in a consulting firm, Multiface, Inc. Previously, he was a Senior Reliability Engineer with Rockwell International in Troy, Michigan. Until 1975, he was a Product Design Engineer at the Ford Motor Company. From 1968 to 1972, he was a Teaching Associate at the Michigan Technological University in Houghton, Michigan where he earned his PhD.

A senior member of ASQC, he is certified by them both as a Quality and Reliability Engineer. He is a Registered Professional Engineer in Michigan and in California. He is a member of ASME, SAE and NSPE.

Quality & Productivity—Have Your Cake & Eat It Too!

C.R. Edmonds
Manager, Technical Systems Planning & Control
Xerox Corporation
Joseph C. Wilson Center for Technology
Webster, New York 14580

Quality & productivity are becoming compatible siblings, not rivals in the Xerox Manufacturing Division. The key to this turnabout is the advent of relatively inexpensive minicomputers. The purpose of this paper is to share some of our experiences both good and bad in the use of inexpensive computers for process control in a labor intensive manufacturing environment. As the title implies, a primary objective is to demonstrate from actual experience that Quality & Productivity need not be mutually exclusive goals.

In the traditional manufacturing environment the only check and balance to give "make it right the first time" meaning was the human being. This included the operator, inspector, quality control engineer, foreman, and production control expeditor to name a few. As a result, with many organizations and many people interacting, each with his own information, opinions, and goals, the interpretation of the facts varied. Also, the information tended to be anywhere from a day to a week old which contributed greatly to differing interpretation. As a result, Quality Control was often perceived as a major roadblock, striving for the impossible (ie - zero defects), standing in the way of making productivity targets.

Today, the minicomputer is becoming the impartial watchdog which stores and analyzes every piece of information it receives. The information tends to be minutes to hours old; a quantum jump in timeliness of information on which decisions are based. The same people make the same decisions, but the quality of the decision has improved as more information is available in a more accessible and timely manner. Computer monitoring of responsibility and delinquency, with exception reporting to the responsible function, insures timely identification of problems and delays without manual follow-up. As a result of the improved checks and balances achieved through process control use of minicomputers, Xerox has observed significant progress toward simultaneous improvement of quality level and productivity.

In the Manufacturing Division the tracking of acceptable or nonconforming material using a minicomputer controlled network of terminals identifies responsibility in real time. With timely knowledge to all of actual status, the time lag of work-in-process material flow has decreased significantly. Correspondingly the frequency of line stoppers has also decreased dramatically. As a result, inventory levels have been reduced and overall productivity has improved.

Using minicomputers the defect data is stored in an on-line data base. This becomes the basis to inspect where needed and reduce inspection where there has been no past problems. This has decreased inspection time, and more significantly, improved throughput. In an assembly line, random defects can be fed back immediately to the earlier source of the defect to minimize recurrence and initiate corrective action. A logical extension of this capability on an assembly line is mechanization of the what, where, who, how type of instructions, identifying specific inspection characteristics to be checked. By placing this information in a video display the ability to inspect or not inspect a specific characteristic based on current experience is realized for the first time.

DEFINITION OF TERMS

Now let me back up to define some terms that will be often used herein. A minicomputer is a general purpose processor with peripherals such as disk and teletype able to handle several on-line/batch tasks concurrently. A more recent product of computer technology is the microprocessor which is a CPU with core memory, and nothing else. It has no peripherals and is able to handle a single specialized task impressed on it from an outside source. The matrix below summarizes some of the key differences in minicomputers and microprocessors.

COMPUTER CAPABILITY MATRIX

TYPE / CHARACTERISTIC	TYPICAL MFR.	COST	LEASE / PURCH.	AVG. CORE SIZE	SOFTWARE CAPABILITIES
MINICOMPUTER	DATA GENERAL C300 HP3000-II DEC-PDP11/70	$100K-200K DEPENDING ON OPTIONS	LEASE	200KB	FULL OPER SYST. GOOD DBMS/ UTILITIES COBOL/FORTRAN, ETC.
MICROPROCESSOR	INTEL 8080 DEC LSI 11 DG MICRONOVA	$1,000 TO $2,000	PURCHASE	UP TO 64 KB	SOFTWARE EXTREMELY VARIABLE CROSS COMPILE FR. BIGGER SYSTEMS COMMON

From a user standpoint the microprocessor is seen as having an increasingly more important role in process control systems. We have made great progress in automatic control of test/production tools and gages using microprocessors to the point where test/production tools have been consolidated into one fixture. The ease with which test limits can be adjusted with fixture wear is key. As a result centralized calibration is required only for the test piece, the fixture can be easily recalibrated in place. Another immediate benefit is "incontestable documentation of defect nature and magnitude. The "throwaway" microporcessor on a chip is evolving rapidly in the marketplace. By 1980 it will cost in the $10 to $100 range and have the power of an IBM 1401. This capability foretells the day when on-board microprocessors will impact everything we touch from automobiles, to washing machines, machine tools, office copiers or typewriters. It will facilitate precise remote diagnosis of product performance problems. The repairman's toolkit will become more flexible based on knowledge of the problem before he leaves the service center.

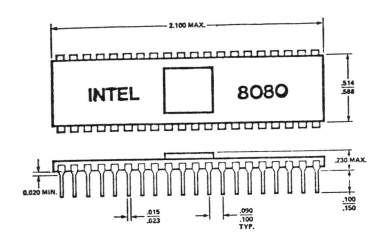

Also, the scope of process control systems includes both control of the machine tool or fixture and the information identifying material flow from those same tools and fixtures. We think of these areas of improved process control as "Islands of Automation". Still deficient, however, is communication between islands to advise of completion or delay of key events to the next operation(s) in the process.

Heretofore, the cost of a totally integrated network of process control computers has been prohibitive except in large volume processes (ie - automotive). However, the day is fast approaching when cost effective computers and associated communications software will be available for lower volume operations. Advances in memory and data storage technology are continuing at a fast pace. Price reductions are projected at a rate of 40% per year thru 1980. (3) Terms such as magnetic bubble, charge coupling domain tip and electron beam access reflect technologies which work in the laboratory.

Our experience indicates a trend toward decentralization of processing. This trend will focus on a "host" minicomputer capability within each manufacturing operation. The host computer would support communications to myriad other more specialized mini/microprocessors used for machine tool or assembly/test fixture control. Another potential use of the "host" minicomputer would be support of numerically controlled machines for program development and down load. The "host" processor would act as a backup in event of failure at the individual processor as well as data accumulation and analysis. On the other hand, the individual processor can continue to operate when the "host" minicomputer fails. The big three in the minicomputer market at present are Digital Equipment, Data General and Hewlett Packard. These companies are moving aggressively to broaden the scope of their products to achieve hardware/software compatibility from the microprocessor up to the super minicomputer class. A key issue is the need for two-way communication between the process control networks and the IBM oriented financial and planning systems. This step is essential to the next quantum jump in timeliness of process control systems thru automated initialization and revision of production schedules. This issue will be resolved with the development of network telecommunication software compatible with the IBM Systems Network Architecture and Synchronous Data Link Control protocols. COADASYL standards are being developed in these areas. The degree of compatibility with those standards will be a key determinant regarding plug-in compatibility of multi-source processors in a distributed network. Another key evolving area in minicomputers is that of Data Base Management Systems. The compatibility and accessibility of data bases across a full range of maxi/mini/microprocessors is essential.

Let me talk now of two relatively new capabilities for use in a manufacturing environment. One tool is bar code scanning of both work-in-process and finished product for configuration control. (4) Is the analogy of the supermarket to a manufacturing environment so far-fetched? High volume or critical material can be more tightly controlled against loss or unnecessary delay. The alphanumeric scanning capability is being developed, which will afford still more flexibility. Consider the ease of a stock purge of obsolete materials or the ease of cycle inventory control with a portable scanner. Also, the portable scanner can be used for periodic configuration/reliability updates of product content at point of use. Consider the potential of product retrofit or reconditioning assessment by taking a portable scanner to the product. All of these benefits would evolve from steps taken to identify selected critical components with a bar coded label during manufacture.

Another new capability is use of computer or operator selectable microforms (ie - slides) as a training and education aide. The ability to display detailed information at each work station on a video screen is a quality improvement breakthrough. Operator ability to display a wiring diagram or exploded view of the assembly being built in the language of his choice is an exciting option. Another use would be computer controlled follow-up of a defect detected by the automated assembly fixture described earlier. Here the automatic selection of a troubleshooting guide would help a repairman walk through a problem.

LIMITATIONS

Heretofore, I have described in positive terms the benefits of mincomputers and microprocessors. Now, let me play "devils advocate" and identify some limitations. Obviously, neither is a panacea. Some of today's minicomputers are running out of steam, placing undue and sometimes unexpected limitations on the user. For example, a computer that barely meets present day needs will probably not be adequate in the near future as minor enhancements are identified. This is especially true of the microprocessor designed into a product or machine; when a need to upgrade occurs, there may simply not be enough unused core or I/O capacity to handle the task. The systems design should always assure field expandability of memory and peripheral capabilities. In fact, this should be a prerequisite to project approval. If it is not, the risks should be identified. Due to technological progress, economic pressure to obsolete a minicomputer or microprocessor will be considerable, until the cost to redo the software is carefully studied. Microprocessors require a very low hardware investment but the task to develop hardware/software interfaces can be highly specialized and expensive. We have found that hardware cost takes up about 20% to 30% of systems development cost. With the microprocessor specifically some unique concerns are evident:

- A tendency to relax backup procedures due to relative invisibility of microprocessor.
- Maintainability concerns (ie - shielding, interchangeability).
- Relative inflexibility. Use where needs are clearcut, repetitive, and major enhancements are unlikely.

IMPLEMENTATION GUIDELINES

The following guidelines may reduce post-implementation problems:

1. Think of computers and systems as if they were building blocks. The evolution of improved process controls is an endless one due to a number of key factors:
   - User perception of what computers can do is growing constantly.
   - The marketplace is very competitive. The cost vs. capability ratio is constantly improving.

2. A backup "host" minicomputer in an on-line process control system is invaluable for batch processing and program development as well.

3. Don't overcomputerize. It isn't worth a $10,000 outlay to mechanize a task which required one hour of a person's time per week.

4. Better to install several small dedicated systems rather than trying to do everything on one large minicomputer. Loading too much on one system increases software cost and limits flexibility.

5. The transition to a small minicomputer system with on-line update can shortcut necessary manual controls and safeguards. On-line edits can't do it all, make sure the system being replaced is thoroughly understood.

SELECTION GUIDELINES

What about selection of a minicomputer manufacturer? How can a potential user make sense out of all the claims and counterclaims?

1. Establish that the vendor can install and service the installation effectively. Only after this concern has been answered should cost and technicalities be considered. User management does not care about technicalities or having the latest gizmo in the marketplace. So far as service is concerned, question duration of downtime every bit as much as frequency. Make sure you talk to several current users of identical or similar product for each vendor being considered. The older installation is preferable for assessment of on-going support.

2. Mixed systems consisting of hardware/software products of several vendors can create serious problems so far as buck passing. If you mix'em; be prepared to maintain yourself.

3. If you must rely on a turnkey vendor for initial development, do so. Everyone has to start someplace.
   - Turnkey systems are like marriages - easy to get into, hard to get out of. The people factor is extremely important. You must feel that you can get along with the vendor's people. User must still fully assess and communicate precise needs. Once those needs are documented in a total specifications package even minor changes tend to become difficult and expensive.
   - Ongoing maintenance or enhancement of existing systems should be handled by in-house personnel. Training of in-house personnel to insure autonomy of maintenance effort should be a key element of initial development contract.

CURRENT APPLICATIONS

Now, let me describe specific existing systems. The systems are by no means all-inclusive, rather they are a cross section selected to show how the approach to solve a problem has evolved in step with the technology. The first three examples are stand-alone component test systems developed in the early 70's.

- The coordinate measuring machine has a DEC PDP8 minicomputer integrated into the system. This is a common turnkey package available from the major manufacturers. The system also includes a teletype device for input/output communications. Major benefits realized by this system are:
    - Improvement in setup time of 66%.
    - Improvement in inspect time of 20%.
    - Minimizes special tooling requirements to inspect complex parts.

- The Cycle Switch Test System has a DEC PDP8 controlling two test fixtures each of which test a number of different switches concurrently. This system uses video displays to show discrepancies found. Some benefits achieved were:
    - Test speed improved 85%.
    - Tests five different parts vs. single part.
    - Excellent repeatability.
    - Virtually no maintenance in 4 years.

- DIGICAM is a test system for inspections of cams. It uses a DEC PDP8 minicomputer with teletype.
    - This system improved processing speed per part from 12 hours (with a manual system) to 7 minutes.
    - Substantial savings realized.
    - Incontestable hard copy record of actual test results is invaluable.

The above systems are ideal examples of the potential for improved control by a host minicomputer. The host would support centralized program development and down load executable programs on request. Also, none of the above have data base or decision making capabilities of their own. As a result of network linkage to a host, overall flexibility and throughput would improve significantly.

The next three applications are process control systems which have evolved in the last four years. Each involves a network of minicomputers and terminals used to measure and manually adjust the process. The systems are described in more detail in reference #1.

- The Photoreceptor Information System is used for on-line inspection and process control. It makes computer aided decisions to inspect or skip/accept or reject. It's heart is the on-line data base of past process experience. The system paid for itself within the first year of operation. The minicomputer driving this system is now obsolete. Replacement with a more up-to-date host minicomputer is underway.

- A second Photoreceptor Information System was a major outgrowth of the prior System. Dual host minicomputers control a network of 14 minicomputers and 22 data terminals. This is an on-line real time system which directs inspect decision making, dispositioning and inventory tracking. The system was necessary due to the complexity of the electrical test specification. In practice a significant improvement in yield has been realized.

- The Work-in-Process Control System is an information handling and statusing tool. It features dual Data Pathing minicomputers and a network of 150 data entry terminals. The system is used for time and attendance as well as tracking work-in-process material flow. The system currently does not have an on-line data base for real time decision making. A response time of 24 hours is adequate to identify process delays. The system has facilitated a 50% decrease in operating cost at the same time improved communications between organizations on the shop floor. This system is also a logical candidate for integration into the host minicomputer network.

The final set of existing systems are those using microprocessors to control individual test fixtures. These were developed in the last two years.

- The 9200 duplicating system final assembly process uses 25 microprocessors controlling individual work station gaging at key points in the final assembly process. These fixtures control dimensional, electrical and optical conformance. The microprocessors used are Intel 8080's tied into Burroughs visual displays with strip printer output.

- The Automatic Data Acquisition System used for control of final assembly operations uses microprocessors to control 16 final assembly stations. Each microprocessor station controls probes at critical locations to gage proper assembly. Each station performs a self-calibration before each test is performed to assure test integrity. Reprogramming of test limits is accomplished via a plug-in switch matrix. The system operates with minimal downtime and has had good operator acceptance.

CONCLUSION

The minicomputer/microprocessor capabilities have had far reaching impact on the Xerox Manufacturing environment. Major progress in process control have taken place with improved quality and productivity. These gains have been driven almost entirely by user organization need. In this environment there will be a continuing enphasis on developing new "Islands of automation" as well as improving what exists. It has been my experience that any process control system, once developed must have continuing support dedicated to its enhancement. The axiom of "Grow or Die" seems especially apropos.

Another major factor in the next five years is the continuing rapid rate of technological development of new computing tools. The effective application of these technologies to solve your process control problems will be largely dependent on developing a nucleus of in-house expertise. However, this growth over the next five years will be more closely controlled in terms of hardware/software guidelines. These guidelines will create a more homogeneous environment conducive to the gradual integration of computers discussed earlier. That integration in itself involves a major learning process for manufacturing management as information, whether favorable or unfavorable, becomes transparent to all. The integration of the "islands" into a total hierarchy may progress slowly, due not only to technological constraints, but due to the transparency problem as well.

The impact of information transparency tends to vary with the organization on the shop floor. The production and operations control segments will tend to be more negative as the information on which their decisions are based has been automated for some time. On the other hand, the quality control segment has often suffered as the information on which a decision was based was not mechanized. Thus, the decision became subjective and judgemental in nature. The trend toward a shop floor hierarchy of computers will cause closer equality of decision making tools between functions on the shop floor. The title of this paper expresses the thought that quality and productivity goals need not be diametrically opposed any longer. With accessible and timely information available to all when a shop floor decision is needed the best interests of all can be served.

REFERENCES

(1) "Minicomputer Application to Quality/Reliability Problems"
C.R. Edmonds/C.W. Hoehn
1976 ASQC Technical Conference Transactions, pp. 352-359

(2) "Basics Often Ignored in Selecting Gear"
Richard A. Kuehn
Computerworld, 5/10/76, pp-7-8

(3) "Memory Prices Seen Dropping 40% per Year until 1980"
J.P. Hebert
Computerworld, 6/7/76, p. 57

(4) "Bar Codes for Data Entry"
E.K. Yasaki
Datamation, 5/75, pp. 63-68

On an Axiomatic Approach to Manufacturing and Manufacturing Systems

by N.P. Suh
Professor of Mechanical Engineering
Director, Laboratory for Manufacturing and Productivity

A.C. Bell
Assoc. Professor of Mechanical Engineering

D.C. Gossard
Assist. Professor of Mechanical Engineering
Massachusetts Institute of Technology,
Cambridge, Mass.

Introduction

A manufacturing system comprises a large number of distinct processes or stages which, individually and collectively, affect the productivity of the overall system. Among the most important of these processes are design, materials selection and processing, assembly and quality control. The interactions between these various facets of manufacturing system are complex, and decisions made concerning one stage of the realization of a product have ramifications which extend to other components.

As such, improvement of the productivity of a manufacturing system constitutes an optimization problem. For maximum productivity, the entire system of manufacturing must be optimized. While optimization of individual system components may produce incremental productivity improvements in system productivity, there is question as to whether improving the effectiveness of individual components can result in the optimization of the total system. For example, the most effective way of increasing the productivity associated with drilling and tapping holes in a finished part may be through the elimination of the need for the holes, rather than through optimization of drilling and tapping operations. Optimization should not be applied solely and independently to parts of the overall process, but rather must be applied to the entire process to maximize system productivity.

Optimization of the complete manufacturing system, however, is not currently possible because we do not have the knowledge base which completely and accurately describes the complex interactions between the various facets of manufacturing. We have, instead, a non-systematic, somewhat random method in which design engineers, manufacturing engineers and others get together, exchange views and formulate a set of decisions which ultimately result in a product. At the present time, most decisions are made to optimize the components of the manufacturing system rather than the total system, because there is no systematic technique which can be used to make "correct" decisions, i.e., those which maximize overall system productivity, at each stage of realization of the product. The degree of "global" optimization now depends solely upon the ingenuity and capability of industrial engineers. The result is often less than optimal and results in low overall productivity.

Global optimization, of the type described above, may be accomplished by two different approaches. The first approach requires a complete and thorough understanding of the complex interactions between the various facets of manufacturing. This understanding would permit an exhaustive evaluation of the effect of each decision on overall system productivity. This might be called an "algorithmic approach" and is the focus of a considerable body of current and past research.

The difficulty with this approach is the global understanding required of an incredibly large data base. Starr (1), for example, describes the need for objective decision procedures in his text but quickly illustrates that a decision tree for even a small number of components and characteristics explodes in the number of options to be considered. There have been numerous attempts to systematize the design of products and processes for manufacturing. Many designers like Blake (2) feel that a systematic approach restricts creativity, but others like Harrisberger (3) outline creative methodologies. Some of these are relatively informal, but the morphological approach, for example, has been extensively studied (4,5,6). All share the feature of generating a large number of alternatives for subsequent evaluation and analysis.

This paper discusses an alternative approach to maximizing productivity of the entire manufacturing system. Called an "axiomatic approach", it is based upon a bold hypothesis:

> There exists a small set of global principles, or axioms, which can be applied to decisions made throughout the synthesis of a manufacturing system. These axioms constitute guidelines or decision rules which lead to "correct" decisions, i.e., those which maximize the productivity of the total manufacturing system in all cases.

Definition of Terms

The purpose of this paper is to explore an axiomatic approach which may eventually contribute a methodology for optimization of the manufacturing system. The essence of the question addressed here is: "Given a set of functional requirements for a given product, are there generally applicable axioms which yield correct decisions in each step of manufacturing (i.e., starting from the design stage to the final assembly and inspection stages) so as to devise an optimal manufacturing system?"

The term "axiom" is used here in the same general sense as thermo-dynamic axioms, which provide basic guidelines to the study of thermodynamics. Specifically, an axiom is a proposition which is assumed to be true without proof for the sake of studying the consequences that follow from it. Manufacturing axioms should provide reference guidelines, which force decisions toward optimization of the entire manufacturing system when followed. Strictly speaking, an axiom must be a general truth, that is, a rule for which no exceptions or counter-examples can be observed. Similarly, an axiom cannot be proven, rather it must be assumed to be true until a violation or counter-example can be found. An axiom must be general; for a manufacturing axiom to be useful, it must be applicable to the full range of manufacturing decisions. By implications, there should be a relatively small number of manufacturing axioms.

The term "corollary" is used here also in the mathematical sense. A corollary is an immediate or easily drawn consequence of an axiom or set of axioms. In contrast to axioms, corollaries may pertain to the entire manufacturing system, or may concern only a part of the manufacturing system.

Description of the Axiomatic Approach to Manufacturing

The first step in the axiomatic approach to "optimization" of a manufacturing system is the specification of the "functional requirements" of the end product. The determination of functional requirements is discussed in Ellinger (7) and Glegg (8) at some length. Functional requirements are defined here as a "minimum set of independent specifications" that completely define the problem. Examples of functional requirements are: kinematic load requirements, expected life under a given set of temperatures, pressures and environment, efficiency, input power, etc. Functional requirements can be ordered in a hierarchical structure, starting from the primary functional requirement to the functional requirement of least importance.

In addition to these functional requirements, there may be the need to specify "constraints". Constraints are defined as those factors which establish the boundaries on acceptable solutions. Constraints on the product may be in the form of acceptable cost, OSHA requirements or adaptability to existing systems. The difference between functional requirements and constraints is that functional requirements are negotiable final characteristics of a product, while constraints are not.

Once functional requirements and constraints are specified for a given product, the design of the product can proceed conceptually. During each stage of realization of the product, axioms can be used to make decisions. Each decision must be guided by axioms and their corollaries and must not violate them. A product designed by following the axioms should yield a design which can be made more productively than otherwise. Similarly, the functional requirements and constraints may be specified for a manufacturing process and the manufacturing process may be synthesized following a set of axioms, again yielding a maximum productivity for that specified process.

A Methodology for Developing Manufacturing Axioms

As described earlier, axioms have two fundamental characteristics:
1. They cannot be proven.
2. They are general truths; no violations or counter-examples can be observed.

These characteristics naturally suggest a heuristic approach to the development of axioms. The heuristic approach involves positing an initial set of axioms. Untested and untried, these "hypothetical" axioms can then be subjected to trial and evaluation in manufacturing case studies. The extent to which these hypothetical axioms satisfy the requirements for true axioms can be assessed. The evaluation can be used to expand, redefine and refine the original set of axioms, until the process converges on a comprehensive set of axioms.

Hypothetical Axioms

To begin an axiomatic approach, a starting set of axioms must be stated. They are intended for use in the design of a product <u>and</u> its processing and production. They are therefore stated as directives rather than as observations.

Axiom 1. Minimize the number of functional requirements and constraints.

Axiom 2. Satisfy the primary functional requirement first. Satisfy the others in order of importance.

Axiom 3. Minimize information content.

Axiom 4. Decouple or separate parts or aspects of a solution if functional requirements are coupled or become interdependent in the designs or processes proposed.

Axiom 5. Integrate functional requirements in a single part or solution if they can be independently satisfied in the proposed solution.

Axiom 6. Everything being equal, conserve materials.

Axiom 7. There may be several optimum solutions.

Axiom 1 calls for a minimization of functional requirements and constraints on the system and guides designers and process planners away from attempts at solving too many problems with one design. It forces constant re-evaluation of the functional requirements and constraints imposed on a problem and avoids the common fault of overspecification so typical, for example, of military purchases. Productivity can always be increased if a product is relieved of some of its requirements and constraints without any loss to those remaining.

The second of these axioms calls for satisfaction of primary requirements first and requires careful selection and ranking of the functional requirements. It mitigates against a common design error, that of losing sight of the primary objective in pursuit of details. Also, it often occurs that a desired set of functional requirements cannot be met or that they conflict in some unsuspected way. This axiom assures a clear decision: choose those requirements with highest priority.

Axiom 3 requires minimizing information and needs an interpretation as to what is meant by "information". The axiom uses the word in the most general sense as the instructions necessary to describe the parts of a product, the processes for making them and the procedures for assembling them. Information refers to both the geometric and surface details of the parts <u>and</u> to the spatial information necessary to move, locate and assemble them. Minimizing information means loosening tolerances, simplifying shapes, accepting rough surfaces. It also implies that the part or the process be designed with a minimum number of instructions for locating the part in processing or assembly and that the location information be preserved where possible in processing.

The productivity advantages of reducing the complexity of parts and the precision required in their production is obvious. This axiom also includes handling of parts, however. For example, a large piston had three grooves near its crown for its compression rings and one near the skirt for an oil scraper ring. A significant improvement in productivity was realized by the simple expedient of moving the scraper ring groove higher on the piston walk to a position where it could be machined without turning the piston around in the lathe. There was no change in the information required for part description. The information required to set up the part for machining was reduced by a factor or two, however: only one setup instead of two.

Axiom 4 requires that if a part or process or assembly must satisfy more than one of the functional requirements, independent control of the requirements must be maintained. Otherwise, the functions must be decoupled by separating parts or processes. An example is the plasticating extruder for polymers. It uses a screw both to melt by shear deformation and to pressurize the plastic and thus control the flow rate. Independent control of the pressure and temperature is lost. Consequently, the production time lost in startup of an extrusion line is large because an extensive trial and error period is required to reach steady state operating conditions. Axiom 4 requires separation of the melting and pressurized delivery functions into separate parts of the machine.

Another example is the electric typewriter. It has been successful because it decouples the functions of force control and key selection. In a manual typewriter, key pressure controlled print quality, which was coupled to key selection by keyboard location and finger strength.

A third example is the Reaction Injection Molding (RIM) Machine. These machines force the two extremely viscous components of polyurethane through opposed nozzles into a mixing chamber so that the impinging jets react as they enter the mold. Good mixing and maintenance of a constant mix ratio require very accurate flow control at high pressure in liquids whose viscosity is highly sensitive to temperature. Conventional RIM machines use very large and expensive positive displacement pumps to develop the required pressure _and_ to perform the metering. Axiom 2 suggests decoupling pumping from metering functions. A very successful RIM machine has been developed which uses accumulators to control shot pressure and a separate metering device to control mixture ratio (9).

Axiom 5 advises combining functions in a part or process whenever the functions are not compromised by the combination. An example is unit body construction in the automobile industry where the styling and structural functions of the body can be successfully combined to eliminate the need for a separate structural frame. Another simple example is a self-tapping, self-locking screw, which combines tapping of a hole with insertion of the threaded fastener and locking it. Note that the first four axioms are satisfied as well. We satisfy the specified functional requirements of tapping, fastening with a screw and locking. None can be eliminated so Axiom 1 is satisfied. The primary functional requirement of fastening is satisfied, Axiom 2. Information is minimized by the elimination of the tap, the separate tapping operation, the lock washer and the assembly of screw and washer, Axiom 3. Axiom 4 is satisfied because the combination does not render the functions interdependent.

Axiom 6 states that all other things being equal, conserve material. The proviso that other things be equal restricts the domain of this axiom to decisions between otherwise equal alternatives in terms of the functional requirements and constraints. In these situations, the solution consuming the least material is best. This axiom is not an important axiom to consider when the value added in processing eliminates the material cost so that conservation has little effect on productivity. The limiting factor in material conservation is the additional processing cost required compared to the value of material saved. This may violate cost constraints.

Axiom 7 states that there will be, in general, more than one good solution for any set of functional requirements and constraints. There is no guarantee of uniqueness of a solution obtained. Also, the existence of a good solution does not rule out finding another.

In order to demonstrate the truth of these axioms, they must be tested and checked against many observations and case studies. Once refined and firmly established with coherent interpretations, they will be powerful tools in the development of a methodology for optimizing a manufacturing system.

Corollaries

A large number of corollaries with more specific applications can be derived from the basic axioms. Eight are derived here for illustrative purposes.

Corollary 1. Part count is not a measure of productivity.

Corollary 2. Cost is not proportional to surface area.

Corollary 3. Minimize the number and complexity of part surfaces.

Corollary 4. If a solution satisfies more independent functional requirements and constraints than were originally imposed, the part or process may be overdesigned.

Corollary 5. A part should be a continuum if energy conduction is important.

Corollary 6. If weaknesses can not be avoided, separate parts.

Corollary 7. If secondary functional requirements can be satisfied without violating primary requirements, then integrate.

Corollary 8. Use standardized or interchangeable parts whenever possible.

The first of these arises out of axioms 4 (decouple to retain independence) and 5 (integrate where independence is maintained). Since axiom 4 increases part count while axiom 5 decreases it while both increase productivity, clearly part count alone contains no information about productivity. Axiom 3 (information) prevents a needless proliferation of parts.

Corollary 2 arises from the axiom constraining information (3) and material (6). Surface area measures neither mass nor information content and thus has little effect on productivity.

Corollary 3 follows from axiom 3, minimize information, and often from axiom 1 (minimize requirements and constrants) when a part is serving too many functions.

Corollary 4 is a consequence of axioms 1 and 7 dealing with minimizing functional requirements and constraints and the plurality of optimum solutions. It states the "no such thing as a free lunch" philosophy: If you are getting more than you need, you are probably paying for it.

Corollary 5 (continuum for energy conduction) results from axiom 5 (integrate). If two parts are to conduct energy in some form (heat, electricity, sound, light, etc.), it is advantageous to make them one part to avoid contact resistance or reduced transmission. Axioms 1 and 2 avoid misuse of this corollary where functional requirements other than energy transmission are concerned.

Corollary 6 on avoiding weakness is derived from axiom 4 (decouple to avoid function dependence). If an 'O'-ring groove provides a sealing function that weakens a structure, dividing the structure at the groove may reduce stress concentration by moving the stress elsewhere. It is assumed that other axioms are not violated in so doing.

Corollary 7 arises from axioms 5 (integrate) and 2 (primary functions first). It advises consideration of those axioms in concert, seeking solutions for the secondary functional requirements among the solutions to the primary requirements.

Corollary 8 advises standardization where possible as an outgrowth of several axioms applied globally to an entire manufacturing operation. It is a long standing practice in the automotive industry and one of the benefits of group technology. Group technology is an attempt to maximize productivity using only axioms 3 and 7. Group technology pays no attention to the other five axioms. Standardization and interchangeability obviously minimize information in many aspects of manufacturing, may integrate requirements and take full advantage of axiom 7 (more than one solution).

Constrast Between Approaches

Having outlined the hypothetical axioms and some corollaries, it is appropriate to compare the approaches to design and manufacturing outlined in the Introduction. Traditionally, manufacturing is often a study of detail. Successful designers of products and processes utilize vast mental catalogs of detailed information on purchasable parts and equipment capabilities. Their creative effort is largely an art, an intuitive balancing of the pros and cons of an infinity of possible, but for the most part not practical, solutions. Most modern efforts to bring order to this mental exercise, so that design of products and manufacturing systems can become more rational, rely on computer-based storage, sifting, selection and presentation of the options and the facts. Group technology and its relatives attempt to bring order to the storage and organization of this mass of detail so that designers and process planners can generalize. This represents an attempt in the algorithmic approach: computerize, pattern recognize and automate.

The axiomatic approach is heuristic and more human oriented. It attempts to bring order to human creativity by stating a few general rules that will always lead to good results and that will so narrow the range of possibilities that the mass of detail to be considered is within the capacity of the designer and planner. As with the laws of themodynamics, human creativity is not eliminated but augmented by simple rules that measure progress. As with thermodynamics, interpretation of the rules is not always simple; in fact, most students find it difficult. The important aspect of the axiomatic approach is that simple guidelines, properly interpreted, offer a way to proceed from the very general to the specific, rather than beginning with the details.

The ultimate rational approach to manufacturing and manufacturing systems probably combines aspects of both the axiomatic and algorithmic approaches. Certainly as long as engineers must be the designers, computer-based approaches will require axiomatic methods to interface the human mind to the excess of data he must deal with. That is, any pure algorithmic approach will not succeed without the powerful insight the axiomatic approach can provide in the optimization of manufacturing productivity.

Conclusions

It is highly plausible that an axiomatic approach can provide a methodology for systematizing and bringing disciplinary features to the manufacturing field. Therefore, an axiomatic approach may provide powerful tools in dealing with issues involved in optimizing manufacturing productivity and play the same role the thermodynamic axioms played in accelerating the advancement of engineering and science.

REFERENCES

1 Starr, M.K., <u>Product Design and Decision Theory</u>, Prentice-Hall, 1963.

2 Blake, I.R., Design Techniques, <u>Engineering Design</u>, T.F. Roylanne, ed., Pergamon Press, 1964.

3 Harrisberger, L., <u>Engineermanship</u>, Brooks/Cole, 1966.

4 Zwicky, F., <u>The Morphological Method of Analysis and Construction</u>, Courant, Anniversary volume, 1948.

5 Norris, K.W., "The Morphological Approach to Engineering Design", Conference on Design Methods, Pergamon Press, 1962.

6 Allen, M.S., <u>Morphological Creativity</u>, Prentice-Hall, 1962.

7 Ellinger, J.H., <u>Design Synthesis</u>, Vol. 1, Wiley, 1968.

8 Glegg, G.L., <u>Design of Design</u>, Cambridge Engineering Publications, Cambridge University Press, 1969.

9 Malguarnera, S.C., and Suh, N.P., "Liquid Injection Molding II, Mechanical Design and Characterization of a RIM Machine", <u>Polymer Engineering and Science</u>, Volume 17, No. 2, February 1977.

CHAPTER 8

QUALITY ASSURANCE AND RELIABILITY

Activities Affecting Quality During Engineering

G.F. Dilworth
Head Mechanical Engineer
Nuclear Steam Generation and Equipment
Division of Engineering Design
Tennessee Valley Authority
Knoxville, Tennessee 37902

This paper discusses how the Tennessee Valley Authority's Division of Engineering Design assures that all engineering, design, and procurement activities affecting quality are carried out. The Division of Engineering Design's responsibilities for quality assurance differ in some ways from a typical architect engineering firm since TVA is its own A/E, constructor, operator, and owner. This paper will also address some of the organizational changes that had to be made within the organization to fully comply with regulatory requirements.

INTRODUCTION

My discussion will cover activities affecting quality in engineering and procurement within the Division of Engineering Design of the Tennessee Valley Authority, but first let me tell you a little bit about TVA.

TVA is a corporate agency of the Federal Government, created by Congress in 1933 as a regional resource development agency and given a reasonable degree of the autonomy and flexibility characteristic of a private corporation. Its Board of Directors, consisting of three members, is appointed by the President, with the consent of the Senate, and reports to him. TVA is an independent agency, not a part of any Federal cabinet department. The President designates one member as Chairman. The Board decides upon major policies, programs, and activities. Appropriations for non-power-related projects are made by Congress after a detailed examination of the proposed TVA budget by the Office of Management and Budget in the Executive Office of the President and by Congressional Appropriations Committees.

The electric power program pays its own way, and is financially separate from other TVA activities. Its operating costs are paid out of revenues from power sales, and its construction costs from revenues and borrowings.

TVA's power service area covers about 80,000 square miles with a 1970 population of nearly 6,000,000. TVA sells power at wholesale to 110 municipal and 50 cooperative electric distribution systems. It also sells power direct to about 50 industries and 10 Federal installations with large or unusual power requirements. Most of the municipal and cooperative systems were organized in the early years of TVA to distribute the large supply of power which resulted from harnessing the Tennessee River. The area in which TVA power is supplied, now limited by law, is about the same now as it was in the mid-1940's. An indication of the growth of the power system is evidenced by power sales in 1954 of 30.0 billion kilowatt-hours, in 1964 of 68.4 billion kilowatt-hours, and in 1974 of 106.1 billion kilowatt-hours.

New generating facilities are added to the TVA system on a schedule designed to provide adequate capacity for meeting projected year-by-year increases in the region's power requirements. Because of the long lead time necessary for planning, designing, and building large power plants--nuclear plants in particular--these projects are being scheduled as much as ten or more years in advance.

Our power program is the largest and most diverse in the United States. We now have 28,204 megawatts of capacity in service. This capacity is made up of 16 percent hydro, 63 percent coal fired, 12 percent nuclear, and 9 percent combustion turbines. In addition to this we have 19,712 megawatts of capacity under design or construction. Fourteen units of this planned capacity are nuclear totaling 18,182 megawatts and four units are pumped storage totaling 1530 megawatts.

Practically all of this operating or planned capacity, 45,823 megawatts out of 47,916, has or is being designed and constructed by our own design and construction forces. Handling a design and construction program of this magnitude has instilled a significant amount of pride within our design organization and we almost felt insulted when the AEC issued the 18 criteria for quality assurance in 10CFR50, Appendix B, and told us we had to comply. Our first reaction was, "surely we already comply."

ORGANIZATION PRIOR TO IMPOSITION OF QUALITY ASSURANCE

TVA's Engineering Design organization has grown through the years to meet the challenge of its rapidly growing power system and is now comparable in size and versatility with many of the large Architect/Engineer firms. At the time of the implementation of 10CFR50, Appendix B, 18 criteria, our Division of Engineering Design was composed of approximately 900 people and organized by discipline into four branches: Architectural, Civil, Electrical, and Mechanical. Each of these branches was responsible for their discipline's share of all of the engineering design and procurement activities for all of TVA's projects.

This organization was put together in such a way to assure a high quality of design. To begin with, there was a high experience factor due to the stability of the organization over a long period of time. Also we incorporated an independent review or check of any engineer's design by someone of more experience plus a review by their common supervisor and then an independent review by a staff specialist, Principal Engineer, who was not in the direct management line. We also used extensively the method of multidiscipline review (squad checking) through all branches. No drawing could be issued until it had been "signed-out" through all branches. One feature of this organization was, in most cases, all functions of engineering, design, and procurement were performed by the same organization. When the QA requirements of 10CFR50, Appendix B, 18 criteria, were imposed, we relied heavily on the Principal Engineer for verification of design and procurement activities. We also had a small QA staff which reported to the Director of Engineering Design. Their main functions were QA coordination with external regulatory bodies and other divisions within TVA and audits of the Division of Engineering Design.

PRESENT ORGANIZATION

It became apparent in the early seventies that the organization by branches according to discipline was becoming unwieldy due to the new projects we had started. The branches were becoming enormous in size and individual project control was becoming more difficult. Therefore, in the summer of 1973 the entire division was recognized as shown in figure 1.

This new organization allowed us to take advantage of experienced personnel in key areas of our engineering branches and to still obtain the much needed independence needed in each project for decision making. It also allowed us to separate the responsibility for activities affecting quality into different organizations and thereby achieving a greater degree of independence. It immediately became apparent that we would have to document in engineering procedures, guides and standards, administrative procedures, and quality assurance standards and procedures how we were to function as an organization which has now grown to 2300 people. The preparation of these standards, guides, and procedures was a massive effort and is now essentially done.

As you can see on figure 1 the Division of Engineering Design (EN DES) is divided into three groups of branches--the Thermal Power Engineering Branches (TPE); the Thermal Power Engineering Design Projects (TPED), and the Special Projects Engineering and Design Branches (SPED). The responsibilities of these three groups of branches are as follows:

   a. The Thermal Power Engineering Branches are primarily responsible for developing the general criteria and basic design parameters for mechanical and electrical equipment and systems and all structural and civil engineering aspects of thermal power generating facilities.

   b. The Thermal Power Engineering Design Projects are responsible for implementing the engineering requirements and criteria developed in the TPE branches into final construction level drawings and documents. They are also responsible for overall management of each respective project.

   c. The Special Projects Engineering and Design Branches are responsible for both engineering and construction level design for nonthermal projects as well as certain site related portions of thermal power projects.

The division also has several staff branches or sections that perform service functions. Two of these that are directly related to quality are the Inspection and Testing Branch and the QA staff (these two have recently been combined into one branch).

The remainder of my discussion will deal with the activities affecting quality performed by the TPE and TPED branches.

ACTIVITIES AFFECTING QUALITY

Thermal Power Engineering Branches

ANSI N45.2.11, Quality Assurance Requirements for the Design of Nuclear Power Stations, defines design input as "those criteria, parameters, bases, or other design requirements upon which detailed final design is based." Our design division is organized so that this part of engineering and design is accomplished in our Thermal Power Engineering Branches. These activities are also reviewed within the same organizations and are also subjected to interdisciplinary review by other branches before being formally approved and issued for use to those branches that perform the "design output."

These activities are those that are involved in determining and specifying the design input requirements listed in table I.

The Thermal Power Engineering Branches also provide the technical procurement function for the division for all material and equipment. The activities involved in this function can be best described in the terms of N45.2.11 as those related to design output or "documents such as drawings, specifications, and other documents defining technical requirements of structures, systems, and components." These drawings and documents are prepared using the design input from other branches within TPE or TPED. These documents are reviewed within the same organization and also given an interdisciplinary review by other groups and branches as required.

The Thermal Power Engineering Branches also perform the external design interface control with design and manufacturing groups from different companies such as the Nuclear Steam Supply System (NSSS) supplier. We do not have as many interfaces to control as would normally be the case in the industry since our Division of Engineering Design is a part of TVA and therefore we are the owner.

Another factor that gives us more assurance of a greater level of review of the design of the plant is the strong influence we maintain over our suppliers. We do a rather detailed review of our supplier's designs and maintain approval rights of most of their drawings and documents.

The review for the adequacy of quality assurance requirements for all procurement documents is done by our division QA staff. They also perform audits of our suppliers.

Thermal Power Engineering Design Project Branches

These branches perform the tasks of "Design Output" and "Final Design" as defined in N45.2.11. The designs are again reviewed and checked within the same organization and given an interdisciplinary review within the project branch. Certain documents and drawings are sent to the TPE branches for design verification before the final design is approved. These design reviews are done considering the list of questions in table II.

CONCLUSION

We feel that we have an organization that maximizes the probability that all final designs are adequate to assure the safe and reliable performance of a nuclear plant that will prevent accidents that could cause undue risk to the health and safety of the public or mitigate the consequences of such accidents if they were to occur; however, we also believe we must be careful that we do not complicate the methods by which we assure ourselves that we have the highest quality of design to the extent the actual quality of the engineering and design will suffer. We can emphasize the procedures and record keeping to such a point that we will substitute paper for logic and intelligence. There is no substitute for the competence of each individual engineer or technician that performs activities related to the design of nuclear power plants.

BIBLIOGRAPHY

ANSI N45.2.13, Quality Assurance Requirements for Control of Procurement of Items and Services for Nuclear Power Plants

ANSI N45.2.11, Quality Assurance Requirements for the Design of Nuclear Power Plants

Title 10, Code of Federal Regulations, Part 50, Appendix B, Criterion III and IV

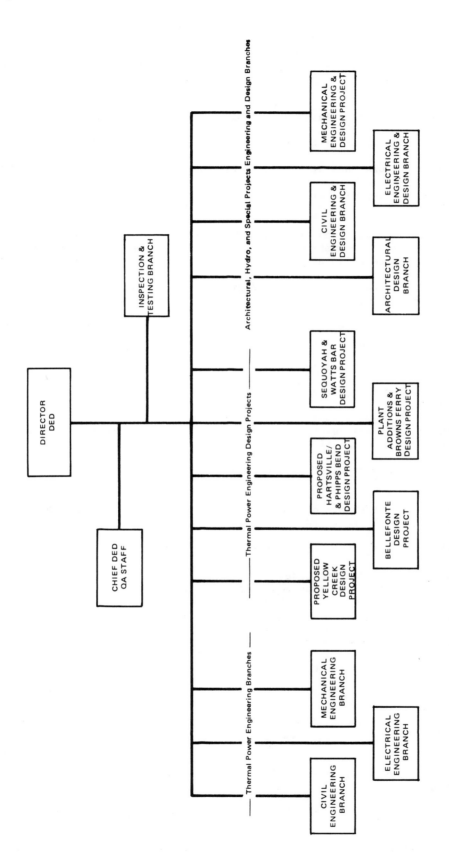

FIGURE 1—ORGANIZATION CHART FOR DIVISION OF ENGINEERING DESIGN

TABLE I

Design Input Requirements

The design input requirements should include the following where applicable:

1. Basic functions of each structure, system, and component.

2. Performance requirements such as capacity, rating, system output.

3. Codes, standards, and regulatory requirements including the applicable issue.

4. Design conditions such as pressure, temperature, fluid chemistry, and voltage.

5. Loads such as seismic, wind, thermal, and dynamic loads imposed by the system itself.

6. Environmental conditions encountered during storage, construction, and operation such as pressure, temperature, humidity, corrosiveness, site elevation, wind direction, nuclear radiation, electromagnetic radiation, and duration of exposure.

7. Interface requirements including definition of the function and physical interfaces involving structures, systems, and components.

8. Material requirements including such items as electrical insulation properties, protective coating, and corrosion resistance.

9. Mechanical requirements such as vibration, stress, shock, and reaction forces.

10. Structural requirements covering such items as equipment foundations and pipe supports.

11. Hydraulic requirements such as pump NPSH, allowable pressure drops, and allowable fluid velocities.

12. Chemistry requirements such as provisions for sampling and limitations on water chemistry.

13. Electrical requirements such as source of power, voltage, raceway requirements, electrical insulation, and motor requirements.

14. Layout and arrangement requirements.

15. Operational requirements under various conditions, such as plant startup, normal plant operation, plant shutdown, plant emergency operation, special or infrequent operation, and system abnormal or emergency operation.

16. Instrumentation and control requirements including indicating instruments, controls, and alarms required for operation, testing, and maintenance. Other requirements such as the type of instrument, installed spares, range of measurement, and location of indication should also be included.

17. Access and administrative control requirements for plant security.

18. Redundancy, diversity, and separation requirements of structures, systems, and components.

19. Failure effects requirements of structures, systems, and components, including a definition of those events and accidents which they must be designed to withstand.

20. Test requirements including in-plant tests and the conditions under which they will be performed.

21. Accessibility, maintenance, repair, and inservice inspection requirements for the plant including the conditions under which these will be performed.

Table I (Continued)

22. Personnel requirements and limitations including the qualification and number of personnel available for plant operation, maintenance, testing and inspection, and permissible personnel radiation exposures for specified areas and conditions.

23. Safety requirements for preventing personnel injury including such items as radiation hazards, restricting the use of dangerous materials, escape provisions from enclosures, and grounding of electrical systems.

24. Transportability requirements such as size and shipping weight, limitations, ICC regulations.

25. Other requirements to prevent undue risk to the health and safety of the public.

26. Handling, storage, and shipping requirements.

TABLE II

Questions to Consider During Design Review

1. Were the inputs correctly selected and incorporated into design?

2. Are assumptions necessary to perform the design activity adequately described and reasonable? Where necessary, are the assumptions identified for subsequent reverifications when the detailed design activities are completed?

3. Are the appropriate quality and quality assurance requirements specified?

4. Are the applicable codes, standards, and regulatory requirements including issue and addenda properly identified and are their requirements for design met?

5. Have applicable construction and operating experience been considered?

6. Have the design interface requirements been satisfied?

7. Was an appropriate design method used?

8. Is the output reasonable compared to inputs?

9. Are the specified parts, equipment, and processes suitable for the required application?

10. Are the specified materials compatible with each other and the design environmental conditions to which the material will be exposed?

11. Have adequate maintenance features and requirements been specified?

12. Are accessibility and other design provisions adequate for performance of needed maintenance and repair?

13. Has adequate accessibility been provided to perform the inservice inspection expected to be required during the plant life?

14. Has the design properly considered radiation exposure to the public and plant personnel?

15. Are the acceptance criteria incorporated in the design documents sufficient to allow verification that design requirements have been satisfactorily accomplished?

16. Have adequate preoperational and subsequent periodic test requirements been appropriately specified?

17. Are adequate handling, storage, cleaning, and shipping requirements specified?

18. Are adequate identification requirements specified?

19. Are requirements for record preparation review, approval, retention, etc., adequately specified?

QUALITY ASSURANCE FOR THE SGHWR*

Quality Assurance in Design

by J.R.D. Jones**
and
P.J. Cameron***

The aim of quality assurance in design is defined in the preceding paper as a disciplined approach to the design of systems, plant and components so as to establish beyond reasonable doubt that they will perform as required and will operate safely and with satisfactory reliability throughout their lives.

Furthermore, it is a means of securing the correct performance of the design task and is not simply a paper demonstration. Good designers are a scarce resource; quality assurance improves their output and can lead to a reduction in the total volume of paperwork generated. It is as important to control the quality of the design input provided by the purchasers as it is to control the output from the suppliers in the form of drawings, calculations, specifications and procedures.

When applying quality control to manufacturing goods the end product is available for inspection and appraisal and the processes which have led to the product can also be well defined. When one applies similar considerations to the design process one needs to examine the design intent (which must be unambiguously stated), the final design and all the intermediate documentation. Additionally one needs to question the thought processes which have led to the production of all these items.

METHOD

This disciplined approach to design conveniently falls under four headings - organization, documentation, verification and audit - and each organization in the pyramid from the utility downwards which has a design function to perform will be expected to show that its arrangements are satisfactory under each of these headings, bearing in mind the importance of its product to the safety and reliability of the station.

Organization

At the start of a design project there will be a number of managerial policy decisions to be made and implemented. As a first step the managerial lines of command, the functions of the various departments and the responsibility of the

---

* Conference organized by British Nuclear Forum and held in London on 13 June, 1975.
** Central Electricity Generating Board, Barnwood.
*** Nuclear Power Company Ltd.

individual design groups must be set down.

This provides the basis on which the policy decisions can be made, as well as ensuring continuing effective administration of the project.

Documentation

A system of documentation must be formulated and operated which defines all aspects of the design at any time and ensures that a logical progression takes place from design intent via design specification to the final issue of the engineering specification. Necessary research and development work must be covered. The final specification will require production of drawings, calculations, manufacturing procedures, materials selection, etc. Amongst the documents which will be required are the enquiry specification, the tender and, in some cases, pre-tender design submissions and the tender assessment. Additionally there will be detailed design documents and drawings, manufacturing drawings, safety documents and commissioning documents, all leading to the final production of the components.

Every design organization should compile formal documents covering those aspects of the design which may be regularized into a standardized form, such as a manual of standards, routines of calculations including the use of specified standard computer codes, design manuals, quality control manuals, and standard plant items and manufacturing processes, etc.

As part of the quality assurance process the key components and calculations, the components with a high degree of novelty and the use of proven designs but in a novel application must be identified. In addition, the results of research projects specifically undertaken to prove a particular aspect of the design must be recorded.

By rationalizing the documentation in this way, the intention is to reduce paperwork and not to increase it.

The process of building up the design in a logical manner also involves detailed programming, the scheduling of design activities and the allocation of adequate resources in all appropriate disciplines and at all levels. The design schedule itself serves not only to remind the designer to carry out specific activities, but, by avoiding the need for him personally to remember these, enables him to devote more time to the solution of the unexpected problems which always arise as the design proceeds.

Finally, there is also a need to keep those people informed who do not have direct access to the main stream of the design evolution but have a vital role to fulfil in the overall design concept, e.g. interacting design groups, safety assessment panels, material engineers, planning and programming engineers, and inspection services, etc.

It is unsatisfactory if the system depends on the initiative and sense of responsibility of individuals engaged on secondary processes to keep themselves informed of the details of the design. It is unreasonable to expect that each such person who may have a contribution to make to the design can anticipate where and how his advice may be useful unless all relevant aspects of the design are made known to him. A clear management control is necessary to ensure the flow of information between these interested parties.

It must be recognized that in any design project it is not possible to progress from the initial design concept to the finalized design without the

the designer having to backtrack on his work as a result of unforeseen problems or changes in requirements.

It should be a maxim that the design documentation represents at any moment the level of completion of the design, so that the technical competence of the design may be verified either as a matter of routine or on a random basis. The design information should come forward in a continuous manner and the design documentation which records and revises this information should similarly be continuous. Whenever changes to the design are made it is important that these should be fully discussed so that their ramifications are understood by all concerned, properly checked and formally approved for issue under the change control procedure, which will be governed by the same control measures as applied to the original design.

Verification

Design control measures must be applied to verify the adequacy of design. The verification should consist of reexamining the design, spot-checking the calculations or analyses and assessing the results against the original design basis and functional requirements.

There are, however, many ways of verifying designs, depending on the importance to safety of the items under consideration and the complexity of the design, the degree of standardization, the state of the art and similarity of the previously proven designs. The proposed method and extent of verification for each item must be documented in an internal design verification plan, made available to the purchaser. The purchaser must indicate in an external design verification plan what additional verification, if any, he intends to impose on his supplier's design. The results of all design verifications must be documented and recorded. The verification process must be performed by individuals or groups other than those who performed the original design, but they may be from the same organization.

Evaluation and Audit

Evaluation and audit may be carried out either by the purchaser or by the quality assurance department of the supplier being assessed. The prime purpose of evaluation is to ensure that a supplier's organization is such that all aspects of quality assurance in design are capable of being carried out effectively. The purpose of an audit is to check that these are being implemented effectively. Design verification, however, seeks to confirm that a particular element of the design is correct.

IMPLEMENTATION BY THE UTILITY AND NUCLEAR POWER COMPANY

The following describes the implementation by the utility and NPC as a system supplier. Implementation by plant suppliers, whilst following the same principles, may differ in detail.

It is envisaged that the utility will write its quality assurance programme requirements into an enquiry specification on its contractors which will become a contractual obligation. It will be the responsibility of each contractor to select, subject to the utility's approval, those parts of quality assurance requirements which are appropriate, and pass these on as contractual requirements to subcontractors who in turn will repeat the process with their suppliers. The organizations, responsibilities and procedures for the production of design input and design output, together with the methods of design verification, will need to be described in the quality assurance programme of each organization.

Within NPC the Water Reactor Division reports back to the NPC Board of Directors. An independent Quality Assurance Unit also reports directly to the NPC Board via a Director with particular responsibility for quality assurance matters. This arrangement ensures that any defect in the disciplined approach to design which becomes apparent to the Quality Assurance Unit is reported directly to the NPC Board so that corrective action can be taken at an effective level in the command chain. Again, Water Reactor Division in NPC is split into two branches, 'technical' and 'engineering', as shown in Fig. 1. The former is responsible for system design, operational aspects, safety and progress monitoring; the latter is in charge of all aspects of plant design and layout. The responsibilities of each chief engineer and group leader within these branches have also been laid down.

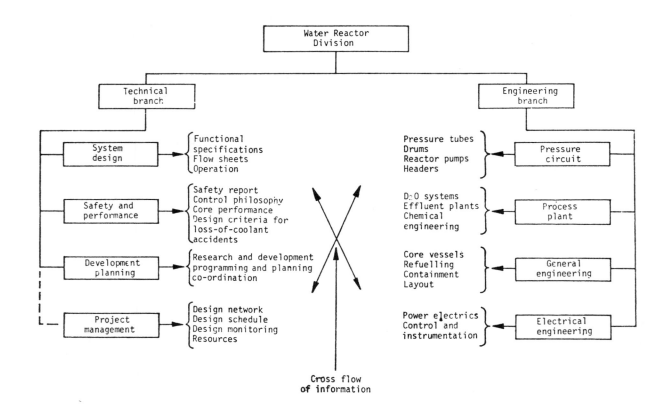

Figure 1

There is deliberate overlap between the responsibilities of the two independent branches and it is the intention that each shall carry out verifications upon the work of the other. Independent verification by the utility will also be carried out.

Within NPC, control documents will be at three levels: project, system and job. Two project control documents are intended: the enquiry specification which will be drawn up by the utility and the project technical specification which will be prepared by NPC. The former will contain all the utility's technical requirements, contract terminating points, site conditions, environ-

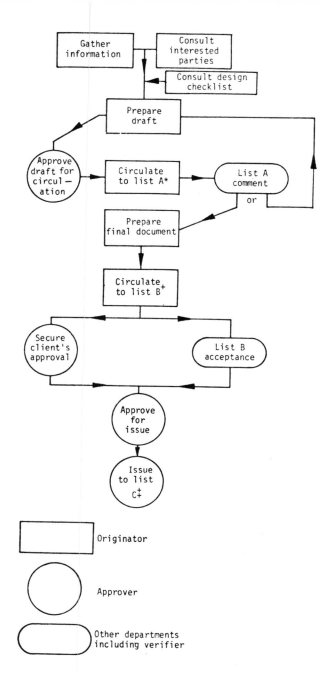

* List A contains all appropriate departments (one copy each) including the client and the appointed verifying engineer.

+ List B includes only addressees whose acceptance is mandatory before publication.

‡ List C provides copies in working quantities to all departments, and the client and the contractor where appropriate.

Figure 2
Procedure for the preparation of project documents

mental constraints, loading schedules, standards, design codes, statutory requirements, etc. The project technical specification, written by NPC and approved by the utility prior to issue, will comprise a number of documents saying how NPC will fulfill the utility's specification. It will be updated continuously.

For each system (defined as a separately commissioned item or group of items) a system functional specification and a system engineering specification will be published by NPC after approval by the utility. Each document will be continuously updated by jointly agreed amendments. The system functional specification will cover relevant input parameters, duty, output requirements, operation, control, safety, maintenance facilities, and outline test and commissioning requirements. The system engineering specification will describe how the functional specification will be met and will outline the reasoning behind the intended solutions.

For each job (defined as a separately procured item, group of items or service) a job enquiry specification and a job contract specification will be prepared. The enquiry specification, which will be produced by NPC, is the basis on which suppliers will tender. The contract specification is the contractor's approved offer and will be formally amended to incorporate any changes as the project proceeds.

Each of the above documents will be prepared in line with the procedure indicated in Fig. 2. Figure 3 shows the front page of a typical NPC system functional specification after it has been through the various stages of approval.

SFS - A 2200

        NUCLEAR POWER COMPANY (RISLEY) LIMITED,
        WARRINGTON ROAD,
        RISLEY, WARRINGTON, CHESHIRE. WA3 6BZ

<u>System Functional Specification</u>

<u>Core Vessels System</u>

<u>660 MW(e) SGHWR</u>

Client Utility 'A'	Station Torwell

		DATE
Prepared by	A.N. Other	21. 3. 75
Approved by	A. Green	23. 3. 75

<u>Distribution</u>:

List C
Contractor (W. Blue) (3 copies)

                    Figure 3

CONCLUSIONS

In essence, quality assurance in design involves no more than formalizing existing good practice, i.e. the setting up of an adequate managerial organization, the rationalization of documentation, the verification of the design, and confirmation that the required procedures are being carried out. The returns are the efficient use of design effort, a reduction in paperwork and a minimum of design errors.

ACKNOWLEDGEMENT

This Paper is published by permission of the Central Electricity Generating Board, the South of Scotland Electricity Board and the Nuclear Power Company.

CHAPTER 9

QUALITY ASSURANCE AND PRODUCT LIABILITY

Quality Assurance And Product Liability

By Richard B. Lynch
Black & Decker Manufacturing Company

A PARABLE

There once were two neighbors who had a hedge which needed to be cut. Because cutting the hedge by hand was tedious, one thought up the idea of cutting the hedge like one would cut grass. There was one slight change--how to hold the mower over the hedge.

No problem, each neighbor would grasp one side and run down the hedge. Fast --easy--simple. Ingenious??

As they were cutting the hedge, they hit a knot. There was then an accident --several fingers were lost.

Who pays? Who is at fault? In this case, the Judge found the use of the lawn mower was blatant, but somebody always pays.

The company won? They had to pay their defense cost and spent $1,500,000 on their product to design out other potential hazards like this in their mowers.

Moral of the story--a manufacturing company must take great care to design, manufacture and sell their product with as limited a liability exposure as possible. Or in other words, always consider the worst misuse of a product and design around it.

I. INTRODUCTION

With the increase today of court cases involving product liability, companies have placed increased responsibility on quality control/assurance areas for expansion into assuring the goodness of the product from the design to the manufacture of a product and, additionally, for its use by the customer.

The purpose of this paper is to focus on the quality assurance function that will go a long way in helping a company limit its liability exposure while coming out possibly with a better designed and "thought out" product.

II. BACKGROUND

A few years ago 50,000 cases seemed to be an extremely large number of cases for product liability. In 1970 we were approaching 500,000 liability cases per year. It is estimated that we will have 1 million cases plus per year in 1975.

The main reason for this increase is additional product liability exposure from the courts. There has been a move towards more liberal interpretation of strict liability and away from privity of contract rules.

Another major reason is that products are getting more complex and more available to a larger segment of the public/consumer. In other words, there are more chances for a product to fail in an unsafe manner or be used in an unsafe manner.

Lastly, the increase in consumerism and the consumer movement has increased both the awareness of the consumer--which is good--and also his inclination to do something relative to faulty products, and accidents resulting from "faulty" products.

It is the job of manufacturers to assure their products are not faulty from a reliability, safety and/or human engineering point of view. This is where Quality Assurance comes in to play a big part.

Today we also have the national commission on product safety with the consumer safety act which can place safety standard requirements on manufacturers. This has and will place additional burden on the manufacturers.

III. QUALITY'S FUNCTION

Management normally sets up a function to serve specific purposes, e.g., production control to plan and control raw material, machines and labor to produce a given product for a given time at the projected cost.

Quality control has been historically given the responsibility to assure that the parts and finish product meet the engineering, marketing and any required regulatory requirements.

Quality control has evolved and expanded over time--more so recently--in responsibility and philosphy to not only the inspection role, e.g., the parts meet the print, but also involving assuring the goodness of design and function to the customer by getting involved early in the design, testing, tooling and initial manufacturing of parts and/or a product.

Quality can be viewed at from the eyes of the user or the manufacturer's. Joseph Juran looks at the "quality" of a product from a user's point of view. He defines quality from its "fitness of use" to the consumer. He defines the fitness of use of a product, material or process as being based on a set of various quality characteristics existing in many areas:

Area	Quality Characteristics
Technological	Hardness, inductance, acidity
Psychological	Taste, beauty, status
Time Orientated	Reliability and servicability
Contractual	Guarantee provision
Ethical	Courtesy of sales personnel, honesty of service shops

A. V. Feigenbaum states that "the word quality does not have the popular meaning of 'best' in any absolute sense." To industry it means "best for certain customer conditions." To the customer quality means that the product functions to his expectations for which he bought it. It is the necessity of having both industry and the consumer understanding of quality to be similar. If a company spends much time and money on a "quality characteristic" which means nothing to the consumer but pays no attention to one that does, they have different understanding.

Quality Assurance, therefore, is a function that management has charged with the responsibility to assure the quality from the customer's needs and wishes, assure the product meets the many quality criterias and lastly, assure that the customer is then told what to expect before he buys and/or uses the product through advertisement, instruction books and other literature.

IV. HOW SHOULD WE ASSURE QUALITY?

PHILOSOPHY

At Black & Decker we approach Quality Assurance from a total quality control point of view. It starts with the inputs at the concept of a unit relative to reliability, function and safety. The underlying principle of our view of total quality--and its basic difference from all other concepts--is that control starts with the initial design of the part and/or product and only ends when the product has been placed in the hands of the consumer--<u>who remains satisfied</u>.

Again, to come back to the topic we are discussing--Quality Assurance relative to product liability, it is for manufacturers to do their homework which will result in a safer product and most times a better design product. This can lead to a cleaner design and a lowest cost design.

We should also not overlook the necessity to understand the essential responsibility for quality and how we organize to achieve it. Some of the basic truths that can be stated about quality are:

1. Basic quality responsibility rests in the hands of company top management and quality objectives cannot be divorced from the overall management of the organization.

2. With the size and scope of many companies today, the job of total quality control cannot be effectively pursued with quality responsibilities widely diffused and separated.

3. But creation of a quality control department does not relieve other company personnel of their delegated quality responsibilities for the discharge of which they are best qualified, e.g., design engineer.

ORGANIZING

A basic question can now be looked at--namely, "how do we organize the quality control organization and where do we put it to assure that we have the control needed in the manufacturing of a product?"

Companies vary widely in product and history and markets and personalities. So, too, will it be appropriate for them to vary in their particular adaptations of the basic quality control structure. Where to locate the total quality responsibility on the organization chart has been debated exhaustively. Should it be part of Marketing, of Engineering, of Manufacturing?

Should it report directly to general management? There are crucial questions and every company has to develop its own answers.

It is our contention, however, that in light of today's enviornment and with the advancement of the quality technology, the basic principle in achieving quality effectiveness is by assuring the <u>independence</u> of this function from other functions of the company. Recognition of this principle, coupled with the staffing of the function with professional practitioners, is probably the most effective manner in which top management can focus and assure its overall quality objectives for the company.

It has been frequently said that "quality is everybody's job in a business", but without clear-cut delegation of responsibility, quality can easily become "nobody's business". Top management, therefore, needs to recognize that the many individual responsibilities for quality are most effectively exercised when they are buttressed and serviced by a well organized, genuinely modern management function whose only area of <u>specialization</u> is product quality, whose only area of <u>operation</u> is in the quality control jobs and whose only <u>responsibilities</u> are to be sure that the products marketed are safe and right --and at the <u>right quality cost</u>. This orchestration can best be accomplished when the top quality position reports directly to the General Manager and is at peer level with Marketing, Engineering, Finance, Manufacturing, etc.

By having the top quality position report to the General Manager (one who is responsible for the profit and loss of the business), management can truly enforce the principle of pursuing quality objectives in a priority with the other business objectives. The quality function, in turn, is not only in a better position to integrate the quality activities of the other function, but can also provide an independent "check and balance" system for assuring the company's product integrity. Under this set-up, the quality personnel, with proper management direction, can better relate to the total business objectives without unnecessarily compromising their position of product quality.

Based on the above discussion, a broad-based quality function reporting independent of other functions can generically be orgainzed as follows:

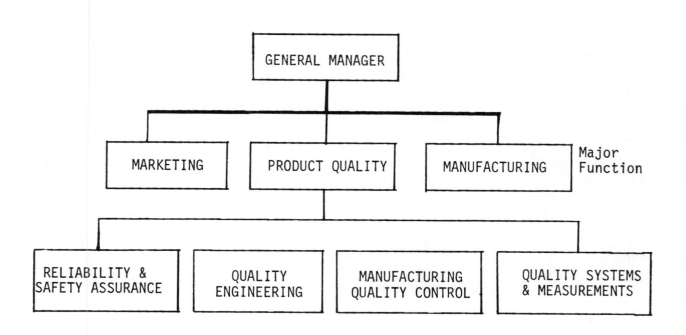

STRUCTURE OF THE QUALITY CONTROL FUNCTION

V. PLAN TO IMPLEMENT QUALITY ASSURANCE

Earlier we stated that the Quality Philosophy is:
Control must start with the design of the product and end only when the product has been placed in the hands of a customer who remains satisfied.

Therefore, in order to have the control needed to assure quality to management, ourselves and to our customers, we must have:
- New design control
- Incoming material control
- Product and process control
- Product quality, reliability and safety assurance control

NEW DESIGN CONTROL

Feigenbaum defines new design control as involving "the establishment and specification of the desirable cost quality, performance-quality and reliability-quality standards for the product, including the elimination or location of possible source of quality troubles before the start of formal production."

New design control's aim is to maximize satisfaction to the customer while minimizing the quality cost. Internal quality cost should be balanced with external quality cost(e.g., warranty cost).

The Quality Engineer at Black & Decker implements new design control through approximately 24 steps:

1. Concept evaluation
2. Product performance sheet
3. Layout evaluation
4. Test program review
5. Safety review 1, 2, 3
6. Critical dimension evaluation
7. Gage review
8. Test equipment review
9. Manufacturing plan evaluation
10. Quality plan
11. Hands on product review
12. Vendor surveys
13. Quality plan documentation
14. Owner's manual review
15. First piece inspection
16. Special studies
17. Field testing
18. Pilot lot report
19. Field test report
20. Final test report
21. Early warning
22. Product release sheet
23. Reliability life testing
24. Packaging review

The Appendix gives a detailed explanation of the points in question on each of the above mentioned elements. Many of the steps happen throughout the development cycle in parallel with other steps. We should also point out that these steps involve the participation of all concerned areas:

- Development Engineering
- Quality
- Marketing
- Manufacturing Plant
- Service
- Engineering Test

Much of the time the Quality Engineer is the "third" person who questions, assists and adds assurance that all needed questions in the development cycle are asked and answered. Depending on the type of product or manufacturing one's company would be producing, many of these steps would be expanded or diminished.

INCOMING MATERIAL CONTROL

Incoming material control is defined by Feigenbaum as "the receiving and stocking, at the most economical levels of quality, of only those parts whose quality conforms to the specification requirements."

This involves not only inspecting the parts or product as received, but assuring that the vendor of the other parts is capable of production with a controlled process so that we can look for trends using statistical analysis techniques.

The initial part of incoming material control comes with reviewing of prints, setting standard and doing surveys on tentative vendors. Once a vendor is selected, part of the purchase contract should include what the vendor will do to assure a product to our specifications and standards.

A quality plan for receiving inspection should be generated that reflects the major and critical characteristics of the part in line with the characteristics and function of the part in use. We must also keep in mind the vendor's processes and control.

Assuring the quality level of incoming material, particularly of the more critical parts or materials will provide a good basis for control of the processes that follow to the final product. This allows a company to limit or eliminate its liability exposure. Here again, we are talking about avoiding product liability: but we will also have a strong basis to defend ourselves with a good receiving control system if a case should arise.

Another point to make is that through receiving records, a company can determine its good vendors from its bad ones. This allows focus to be put where it would be most effective and gives guidance to quality and purchasing on good vendors for future materials, products or processes.

PRODUCT AND PROCESS CONTROL

Product and process control is the control of a product from its initial manufacture, through to the final product utilizing feedback from the processes, fabricating the final product and from input from the field. The main objective is to correct any deviation of a product or part from specification before it becomes defective.

The rewards of good product and process control is minimizing internal and external failure cost and maximizing customer satisfaction, which leads to minimizing the possible liability posture of a company.

The type of control utilized depends on the type of manufacturing done, the inherent control of the process and the risk involved with failure, e.g., scrap cost or chance of a field failure.

We should remember that an inherent and integral ingredient of control is not only the process, but also the operators controlling the process. The operator responsible for a process or machine should have the primary responsibility for making good parts. Quality does not or should not try to be totally responsible for "controlling" quality. This would take away a responsibility that is basic in manufacturing with the manager, supervisor and operator.

Up to now, we have been placing controls on materials and processes to end up with a final product, which we will sell to a customer so that we can afford to continue through the cycle as a healthy business. It is the product quality, reliability and safety assurance areas which:

- Inspects the final product relative to the specification
- Audits the tool for function from the customers point of view
- Life tests products to keep a handle on overall reliability and life of the product
- Lastly, monitors feedback from testing and the field relative to any safety problem that arise or become potential safety hazards (minimize liability exposure)

Final inspection is the final stage where the finish product, in Black & Decker's case, is 100 percent inspected to assure what we are putting out of the door is safe, reliable and meets engineering specifications. It is here that we must be careful to not over-inspect or test, but rather to look for the characteristics that are highly variable and provide the intended function of the product. An example of an electric drill would be: Amperage, speed and voltage leakage.

The final audit should not be used as another inspection station, but rather as a monitor of how the overall quality system is working--keeping an eye on a drift of standards, subjective criteria, etc. Audit will pick up major problems at times--but only because the test area has drifted too far or not looked for the right characteristics. This will not happen if there is a well thought out quality plan, which is used by the tester.

Life testing is a functional test of a product that is subject to wear to monitor the life expectancy of a product from one lot to the next and over a time period. Also, the mode of failure is important from a safety, serviceability and overall reliability of the product.

An area or person in quality should be designated to monitor safety from initial design phases to reports from the field. This provides feedback to an actual or potential safety problem and allows for quick and correct response to any liability cases should they arise.

SUMMARY AND CONCLUSIONS

Quality Assurance does and should play a large role in preventing product liability cases, but it has to be a tops down commitment on the part of a company and its management.

Another point is that quality is still everybody's job. The Quality Assurance area should be assisting the other functional areas in a company to do a good overall job and assure the steps required happen and continue to happen.

Lastly, the quality assurance/control function can be and should be a positive influence relative to providing cost savings through reduction of much internal and external failure costs.

APPENDIX

1. ## CONCEPT EVALUATION

   Visualize the product from the customers standpoint. Look at the needs of design on manufacturing capability. Provide the development group with history and analysis of similar products relative to: Field failure, customer complaint, reliability performance, test bench failure, audit results, accident data, product service, and lastly, all applicable standards.

2. ## PRODUCT PERFORMANCE SHEET

   Here we set down the main specifications or criteria that the tool should meet and identify some competitive tools with which we will test against. With the criteria established: the results from prototype, pilot lot and first product lot are then monitored.

3. ## SAFETY REVIEWS

   Three safety reviews are held to review all safety hazards that could exist-- mechanical, electrical, human factors and miscellaneous. These are attended by Engineering, Quality, Marketing, Manufacturing and Engineering Test to evaluate the tool and its use from all points of view.

4. ## LAYOUT EVALUATION

   Review layout looking at basic design, areas of tight tolerances, identifying need for any special gaging or test equipment, and lastly, reviewing datum system that will best suit all needs.

5. ## CRITICAL DIMENSION REVIEW

   As the design moves from the layout stage to final drawing, review for critical dimensions. Make all parties involved aware of cost effects of any close tolerance.

6. ## TEST PLAN REVIEW

   Review engineering test plan to assure that all safety, functional and reliability parameters are thoroughly tested and proved out in the prototype and pilot lot testing. Make sure reliability testing to be done in the plant duplicates engineering testing.

7. ## GAGE REVIEW

   Review prints for gaging requirements. Hold meeting to propose, review and approve gaging throughout the development cycle.

8. ## TEST EQUIPMENT REVIEW

   Based on the test plan: review, propose, review and approve test equipment to be used in the plant--both final test and reliability testing.

9. **MANUFACTURING PLAN EVALUATION**

    Review processes planned for the parts and finish product. Plan studies required, control points to monitor and first piece timing of purchased and manufactured parts.

10. **QUALITY PLAN**

    Work up quality plans based on classification of characteristics, quality levels required and process capability.

11. **HANDS ON PRODUCT REVIEW**

    Take representatives of all interested areas and use prototype units in the intended manner. Look at the tool from the customers point of view.

12. **VENDOR SUMMARY**

    Visit all proposed vendors, rate them, communicate quality requirements to selected part vendors.

13. **ADVERTISING LITERATURE REVIEW**

    Review all advertising literature for description of tools function, capability and/or limitations from a safety and reliability point of view.

14. **OWNER'S MANUAL REVIEW**

    Review for adequacy of explanation of use and any safety precaution notes required.

15. **FIRST PIECE INSPECTION**

    Do a complete inspection of all characteristics in order to qualify the tooling and/or process on any new parts.

16. **SPECIAL STUDIES**

    Supplemental studies to first piece inspection when process and/or part capability needs to be determined.

17. **FIELD TESTING**

    Determine field test requirements of quality: location and timing. Place units with questionnaires, review feedback and summarize.

18. **PILOT LOT REPORT**

    A report of the complete pilot lot: recording failures, problems, causes and responsibilities. Use as a tool to follow-up all miscellaneous problems to get resolution before production lots start.

19. **FINAL TEST REPORT**

    Review of all testing in Engineering Test of prototype and pilot lot units. Evaluate need of another lot or special studies on Lot 2.

20. **EARLY WARNING REVIEW**

    Review of initial failures from the field: listing quantity, failure, cause and follow-up. Hold meeting with Plant, Engineering, Service and Marketing to review plant reject rates and reliability testing.

21. **PRODUCT RELEASE SHEET**

    Review sheet to check key items completion before recommended release to Marketing by the project team.

22. **RELIABILITY LIFE TESTING**

    Review present life testing method on common existing units. Decide on testing required to assure adequate testing for assuring on going production. Testing should duplicate Engineering Testing that qualified and unit and simulate actual use where possible.

23. **PACKAGING REVIEW**

    Review packaging requirements and drop test performed to assure packaging is sufficient to assure no damage to the product or cause parts to be lost in transit to the customer.

BIBLIOGRAPHY

Feigenbaum, A. V. *Total Quality Control.* New York: McGraw-Hill Book Company. 1961.

Juran, J. M. *Quality Control Handbook.* New York: McGraw-Hill Book Company. 1974.

The Act and its Principal Features. Hartford: The Travelers Insurance Companies. 1972

A Management Guide to Product Quality and Safety. Hartford: The Traveler Insurance Companies; Engineering Division. 1973

Sarin, Lalit K. *Structure of the Quality Control Function,* presented to the machinery and allied products institute products liability council meeting: Washington D.C. 1975.

PRODUCT DEFENSE – A TEAM EFFORT

In these days of the law of strict liability, a plaintiff commonly seeks to prove that a product involved in an accident was defective when it left the manufacturer's plant. If he can successfully attack the product's design, he does not need to prove negligence during the manufacturing process to win his case.

by **C. W. Walton,** Vice President - General Counsel and Secretary, **Koehring Co.**

Large consumer-product manufacturers, such as the automobile companies, have had fairly sophisticated quality assurance programs for many years. The products of some of the smaller companies do not have the same visibility, either because they are components or are not sold to the general public, and often the quality assurance programs of such companies are limited or nonexistent. Whatever it makes or sells, any manufacturing firm should develop an entire-company approach to product safety in these days of large settlements for product liability claims.

Top management

The top levels of the company must clearly recognize the importance of product safety, and set the tone and attitude of the organization. Some companies include a product safety statement in their corporate objectives. Next, they recognize product safety as a separate function reporting to top management—perhaps to the president or an operations or engineering vice president. Without top management endorsement and backing of this program, product safety and product defense will not be a top corporate objective. The time is past when a company can say that safety of its products is one of the many responsibilities of the chief engineer.

A corporate safety program, headed by a corporate product safety manager, typically involves *creation of a product safety committee* consisting of key engineering, marketing, manufacturing, and legal personnel who undertake periodic formal reviews of new and existing products. A checklist or audit form can be used to force the product safety team to consider all aspects of design, guarding, warning, anticipated use, and even foolhardy misuse of the product. The committee should be responsible for reviewing field failures and safety/accident reports, and should make appropriate decisions on recalls or retrofits. The committee must remember that a product liability plaintiff, in these days of the law of strict liability, will usually try to invoke SEC. 402(a) of the Restatement of Torts by proving that a product involved in an accident was defective in design or manufacture when it left the maker's plant. If he can successfully attack the design, he does not need to show negligence in the product's manufacture to win his case.

Engineering

It is obvious that the engineering department should always consider safety in its designs, and try, first, to *eliminate* unsafe conditions: if this cannot be done, to *guard* the unsafe condition: and if neither can be done, to *warn* the user of the hazard by labels and decals affixed to the product and also by warnings in operator or maintenance manuals.

If possible, the engineering department should provide a record of suppliers' approvals of applications of their components to the product in question. Recently a case came to light in which just the opposite was true. A braking system on a truck-mounted construction machine failed, resulting in the death of a woman and several children. When the machine manufacturer went back to the brake system supplier to determine that the system was the proper one, what was actually found instead were several letters from the brake supplier dated years before when the design was being created, warning the machine manufacturer against the particular application and stating that premature brake failure could be the result. The case was settled by a substantial payment by the machine manufacturer's insurance carrier.

Engineering should always determine, by prototype testing, the integrity of new designs. The types of tests and limits of testing which took place are often essential ingredients in subsequent proof of adequate product design.

The engineering department must be well aware of the state of the art of a particular type of equipment, and should know whether any of the competition is able to eliminate a hazard, guard it, or warn against it. The engineer on the witness stand is often asked whether he has knowledge of what the competition is doing —and, if the competition is providing a safer product, then his own product looks bad in the eyes of the jury.

The chief engineer or product engineer may be placed in the role of an expert witness if his product becoms involved in litigation. Some engineers make excellent witnesses: others, by virtue of their training or personality, are poor witnesses. Sometimes they are too contemplative in answering a question, or too willing to state that anything is possible, or they fail to express an opinion unless they have virtually 100% proof of its validity.

Another consideration in choosing expert witnesses is whether to use a witness from the company whose product is involved, a witness from a competitor, or someone from an independent source such as a local engineering firm or university. Again, this may depend on the qualifications and personality of someone within the company as compared with someone outside who may be knowledgeable in the field and make a better courtroom impression. The engineer is usually best qualified to assist counsel in analyzing and attacking the testimony of the plaintiff's expert witnesses.

Manufacturing

The manufacturing division of a company plays a crucial role in the safety and reliability of the end product. It must inspect incoming components and supplies to make certain that these meet specification and quality requirements. Manufacturing must also provide a positive means of identifying the source of critical items. For example, if a fastener fails and causes an accident, can the company determine who provided it? Does the company know which of its own products contain fasteners from the defective batch? If it can be proved that the fastener was defective, recourse may lie with the supplier of the fastener. As equipment or machinery manufacturers become more successful in this type of source identification, the greater is the risk to component and other OEM suppliers for failure of their products.

A quality control function with independent authority to order rework or to prohibit shipment of the company's product is essential whenever the product may not be up to specifications or standards for reliability and safety. Also, critical jobs should be performed only by trained, competent personnel (e.g., only certified welders should make critical welds on construction machinery). Testing of finished products should be a part of the manufacturing process. In conjunction with engineering, certain required tests should be set up, such as lifting capacity, horsepower output, and magnetic particle inspection in highly-stressed areas. In conjunction with the purchasing function, a constant monitoring of the quality of work performed outside in job shops is vital to the safety of the final product. Of course, none of this is of any long-term value unless detailed records are kept to show source of supply, what was tested, how it was tested, and the test results. These records should be carefully preserved in the event that they have to be produced many years later in a lawsuit involving a particular piece of equipment.

Marketing

In the selling area, a firm policy should be established on the offering of safety options. There has lately been an increasing trend in the law to indicate that the manufacturer should include all known safety devices with the equipment when he sells it. This means that safety equipment should not be an "add-on" option. Some companies will delete safety options and deduct their cost from the purchase price only upon specific written instructions from the customer, and require the customer to state in these instructions that he assumes full responsibility to provide equivalent guards, shields, or other safety devices.

The operator and maintenance manuals are extremely important and should be carefully reviewed by the product safety manager, the marketing department, and the legal counsel of the company. A warning or instruction which is ignored but which appears in a manual may mean the difference between winning or losing a lawsuit.

Good maintenance manuals are particularly important because many accidents do not involve normal machine operation, but occur during maintenance or service work. Step-by-step procedures for taking safety precautions when guards or other safety devices are removed will go a long way toward product liability defense.

Specification sheets, brochures, and other printed literature should be carefully scrutinized by engineering and legal personnel. Despite exculpatory clauses in machine warranties, a statement in an advertisement may be considered a warranty and can easily be referred to after an accident has occurred. Representations regarding a machine's safety are particularly touchy, as a plaintiff may try to show that he bought a machine relying upon an advertisement stating that it was a perfectly safe machine to own and operate.

Service files are another important record. One of the key items in the service history of a machine is a set of regular inspection and maintenance reports by the equipment suppliers' field service representatives or a dealer's servicemen. It is important to have good records showing each visit to the equipment, the work performed on it, the condition of the equipment, and a statement of observations of improper use with instructions given to the owner or operator to correct the misuse.

Accident investigation

Despite all precautions, accidents do occur. Prompt investigation of an accident is essential to any subsequent lawsuit. This investigation may be performed by the equipment manufacturer, dealer, or distributor, or sometimes by insurance personnel. Detailed information should be obtained, including a description of the accident, names of victims, witnesses and participants, statements of witnesses, and photographs whenever possible. Accident investigators should be instructed to observe and listen but not to express any opinions. Their job is merely to "freeze the facts." On a report form covering a recent accident, the investigator stated his opinion that the accident was probably caused by an improperly adjusted brake on a hoisting drum. A check of the equipment manufacturer's service reports showed that a service representative had visited the job site during the week preceding the accident and had adjusted the hoist drum brake at that time. The opinion of the accident investigator was fatal to the trial of the case, and the argument of operator mistake was rejected by the jury.

It is also essential that evidence be carefully preserved. If a component fails, or a cable parts, the offending piece should be kept available for investigation, testing, and use as evidence during a trial.

Someone within the corporation must assist the attorney who will try the case in court. This is a liaison function which can be performed by the engineering department, general management, in-house legal counsel, or a combination of these departments. It includes planning of trial defense strategy, evaluating settlement options, and education of counsel regarding the design, manufacture, and use of the product. Most attorneys defending a case are not initially familiar with the product in question. Bringing the lawyer into the plant to observe manufacturing processes and perhaps even asking him to operate the equipment may pay big dividends in imparting a good working knowledge of the product he is defending. If someone within the company is to be the expert witness in the case, he and the attorney should become acquainted prior to the trial date.

Post accident

Whether a case is won or lost, the company should stand back and take a look at the experience. Did the verdict indicate that a general recall of the product may be needed? If so, was it recalled, and how effectively? Should the product be redesigned or retrofitted to prevent similar accidents? Should a notice be sent to all distributors and users of the machinery? Should new warnings be added on the machines or in the manuals?

In many accident lawsuits it can be legitimately argued that the misuse of the equipment which caused the accident could not be reasonably foreseen, and therefore there was no obligation to guard against it. However, when one accident has occurred, a second accident of the same type becomes foreseeable. Too many companies have a tendency to turn over accident cases to their lawyers or insurance carriers and then forget about the situation which caused the accident to occur in the first place.

Remember, *every accident should be a lesson*—it should teach something. ▲

Preparing Tomorrow's Defense Today

Norman G. Sade
Porzio, Bromberg & Newman
Morristown, New Jersey

Good manufacturing practices can enhance the reliability of your product and at the same time, reduce the risk that your company will be held liable in a products liability suit. These practices include quality control, accident consciousness, good record-keeping, design integrity, careful supplier selection, product reevaluation, to name only a few. The failure of the manufacturer to focus on these and other related practices will have profound consequences, as suggested by the program title.

There are two broad categories of products liability cases. These are simply manufacturing defects and design defects.

By way of definition, a "manufacturing defect" is one which results from a failure occurring during the actual manufacturing process - an improperly tightened lug nut, an inadvertently omitted safety shield or an inadequate bolt used mistakenly to secure a vital function. The result of this type of defect is a product which may be dangerous because it does not conform to original design specifications.

Defects occurring as a result of negligent design present a vastly different situation. There are two types of negligent design cases. The first of these involves a product which is manufactured in accordance with the design specifications, but the design itself is faulty. The responsibility would normally fall on the shoulders of the engineers who, in most cases, could have detected the design fault at an early stage by careful and critical design analysis and testing. In the eyes of a jury, a defect due to negligent design may well conjure up the image of the careless corporate giant, making the defense of such a case extremely difficult.

The second type of design case involves the defect arising from the proper execution of a design consciously and knowingly adopted by a manufacturer. A conscious design choice is made with a full awareness on the part of the manufacturer that certain risks of injury are present in the design. The conscious design choice is the product of a balancing of several factors: risk of injury, product utility, design alternatives and economic factors.

Perhaps the best way to illustrate how difficult it may be for the manufacturer in certain cases to make a viable defense in products liability cases and how easy it may be for the plaintiff to prove that there is a defect in the hands of the manufacturer is to consider the current state of the law in New Jersey. Other states may have comparable decisions.

Initially, a product is defective if it is not fit for the ordinary purposes for which such articles are sold and used.[1] To show a defective condition all that is required is:

> proof, in a general sense and as understood by a layman, that 'something was wrong' with the product.... [T]he mere occurrence of an accident is not sufficient to establish that the product was not fit for ordinary purposes. However... proof of proper use, handling or operation of the product and the nature of the malfunction, may be enough to satisfy the requirement that something was wrong with it.[2]

Once a defect is shown either directly or circumstantially, to comply with the prerequisites for strict liability in tort, the plaintiff must show that the defect existed while in the hands of the manufacturer.[3] This too can be accomplished through circumstantial proof based on the nature of the defect or the age of the product. Thus, in a recent case, a plaintiff was entitled to recover for injuries sustained in a collision allegedly caused by a steering mechanism failure in a six-month old Lincoln with 11,000 miles.[4] In effect, the jury was permitted to infer both that there was a defect in the steering mechanism by reason of its failure and that the defective condition existed in the hands of the manufacturer by reason of the vehicle's age. Where circumstantial evidence is used as presumptive proof of a manufacturing defect, the plaintiff, of course, has to negate other possible causes for the accident for which the manufacturer might not be responsible.[5]

What can be drawn from these rules of law is that with respect to a manufacturing defect in a relatively new product, the manufacturer has a heavy yoke to bear. Confronted with such onerous authorities, what can a manufacturer do to meet the challenge of products' liability litigation?

Initially, it is necessary to anticipate the defense of products liability cases. As one commentator aptly stated:

> Making now the best legal case out of the design and manufacturing events of the past simply will not do the job. Rather, the proper defense of products in strict liability in tort suits is to shape the design and manufacturing events of the present so that the best legal case will be available in the future.[6]

A manufacturer who expects to successfully defend a products liability action predicated on a misadventure in the assembly process should have a rigorous and continuous quality control program and the means to prove its existence and enforcement. The quality control program should involve inspection of the product at various points during the assembly phase as well as inspection of the finished product in a thorough and systematic manner. The means to prove the existence and enforcement of the quality control program should be embodied in detailed, complete records. If you are not presently concerned with quality control and its documentation, tomorrow's legal case will be severely hampered - to say nothing of the quality of your product!

Preparing tomorrow's product defense today requires close and constant contact between management and legal personnel. For this purpose it is becoming more and more common for manufacturers to establish departments or committees whose primary function is to act as liaison between the management, including production and engineering, and the lawyer, in-house or outside counsel, charged with responsibility for monitoring or defending products liability cases. The intercourse between these groups permits speedy appraisal of product failures and early response to the existence of claims. Production and design changes are often made in response to information supplied by these groups since products liability suits often focus on weaknesses in the manufacture or design of the product.

The responsibility for drafting adequate warnings and labels, for the supervision of advertising content and for the language of instruction manuals often falls into the bailiwick of these departments or committees.

Not the least of their functions is the responsibility for assisting outside counsel in the preparation of the defense of pending suits. In this capacity they are called upon to investigate data relating to the manufacture of the product, including the design history of the product, the existence of similar claims, and a host of related matters. They are, therefore, in a unique position to obtain information with respect to the product and to guide the manufacturer in reevaluating the acceptability of the design as well as manufacturing aspects of the product.

If there is no individual or group of persons responsible for these functions, not only will the defense of products liability cases suffer for lack of adequate attention but, more important, the manufacturer will suffer from a lack of information and ultimately from a lack of sensitivity to what the product is doing in the market place.

In many cases a manufacturer incorporates into his product component parts supplied by others. Although under some circumstances the manufacturer may have a common law right of indemnification against the supplier, there are some practical steps that can be taken to minimize the manufacturer's exposure to liability. The most obvious of these is for the manufacturer to satisfy himself that the supplier is reliable and that the component being purchased is of acceptable quality. Depending on the nature and complexity of the component, inspection and testing may be required.

The manufacturer is also well advised to purchase no components from a supplier who insists on extracting a release or indemnification agreement from the manufacturer. The manufacturer, on the other hand, should require proof that the supplier maintains adequate products liability insurance so that if the manufacturer is entitled to indemnification in a given case, the right to recover from the supplier will be a meaningful one, and, if possible, he should obtain an express indemnification undertaking from the supplier.

In addition to the considerations mentioned above, design cases involve other rules of law and other concerns for the manufacturer.

Normally, manufacturers are guided by familiar industry standards, governmental regulations and, often, competitors' designs. Although objective standards of safe design are normally admissible in evidence as representing "a consensus of opinion carrying the approval of a significant segment of an industry,"[7] mere compliance with these guidelines will not necessarily result in a safe and defensible product. More may be required. This added element is sometimes called "accident consciousness." This means designing a product which not only accords with existing standards but which is designed with a view towards minimizing injuries which may result from the use of the product.

A recent noteworthy United States District Court case in New Jersey is instructive.[8] Plaintiff's decedent was in the driver's seat of a new 1970 Chevrolet Nova when his vehicle was struck in the rear by another vehicle. Upon impact, his head struck the head restraint (supplied as standard equipment by virtue of government regulations), provided to prevent whiplash injuries. The impact was severe; he suffered extensive brain damage and he died shortly thereafter. Plaintiff's estate charged that the head restraint, consisting of a horizontal piece of metal covered by a foam cushion, was dangerous and that it caused the injury. In response, the defendant attempted unsuccessfully to argue compliance with Federal standards as a defense. The court adopted the rationale of Larsen v. General Motors Corp.,[9] holding that defendant here was obligated to foresee the probability of a second-impact injury resulting from a passenger's head impacting on the head restraint. As to the defense of compliance with Federal standards, the court dryly observed that a "suspended machete blade" would have complied with the standards and probably caused similar injury to the driver. It follows, therefore, that mere compliance with existing governmental standards may be inadequate as a defense to a products liability claim.

Preparing tomorrow's defense today is absolutely necessary in cases involving conscious design choices. If the manufacturer does not undergo a critical design analysis and also maintain records of the considerations underlying design decisions, justifying the integrity of the design in the courtroom will be an extremely difficult undertaking, particularly if the designers of the product are no longer available to testify.

A 1971 Lousianna case is illustrative.[10] There, while the plaintiff was cutting her lawn with a riding mower, she attempted to avoid an obtruction. In reversing the machine, it reared up and threw her directly into the path of the machine, causing severe injury to her foot. Plaintiff alleged that the mower was defective since it lacked a "dead man control" and a detente plate to prevent accidental gear slippage. The manufacturer was able to show not only that the machine conformed to government standards, but that extensive tests were made on the prototype of the machine, that "dead man controls" were originally installed on prototypes but later rejected for safety reasons and that the detente plate was unnecessary. Thus, the manufacturer was able to meet the plaintiff's claims of defect because in its original design analysis the manufacturer considered, reviewed, and for acceptable reasons, rejected the design proposed by plaintiff's experts.

In arriving at a particular design, a manufacturer should go through a painstaking and methodical process of evaluating a number of important factors:

1. What are the intended uses of the product?
2. What uses are foreseeable but unintended?
3. Are there viable alternatives which pose lesser risks of harm?
4. What is the social utility of the product?
5. What safety devices are appropriate?
6. What warnings are required?

If this analysis is undertaken and acceptable answers reached, the manufacturer should be able to successfully defend the product in court.

Intimately associated with the design of a product is the duty to provide adequate safety devices and warnings where necessary. New Jersey's landmark Bexiga and Finnegan cases are instructive as to the manufacturer's duty to provide safety devices.[11] In those cases, decided in 1972, the Supreme Court of New Jersey held that manufacturers of potentially dangerous industrial equipment have a non-delegable duty to install safety devices to protect operators even though, at the time those cases were decided, it was accepted trade practice for the purchaser to install the safety devices.

The public interest in assuring that safety devices are installed demands more from the manufacturer than to permit him to leave such a critical phase of his manufacturing process to the haphazard conduct of the ultimate purchaser. The only way to be certain that such devices will be installed on all machines--which clearly the public interest requires--is to place the duty on the manufacturer where it is feasible for him to do so.[12]

The lesson to be learned from Bexiga is clear: you may not safely delegate the installation of safety devices on your product.

It has been aptly stated that "[a]n adequate warning ... makes reasonable those dangerous defects or conditions in a product that the safety devices cannot cure."[13] Although a warning may not cure all defects in the product, the duty to warn of any known danger is manifest. What can occur as the result of a failure to warn is demonstrated in the following New Jersey case.[14] Plaintiff was employed as a packer, using a machine to tighten, knot, and cut wire used for securing packages. The accident occurred when the machine tightened the wire beyond its tensile strength, causing the wire to snap in plaintiff's face. The manufacturer was found liable for failing to warn the employee by suitable notice on the machine of the necessity of wearing safety glasses even though the employer had been given such a warning orally by the manufacturer but had failed to pass it on to the employee. A simple lable, adequately and forcefully phrased, would have avoided the result.

A warning must, of course, be adequate. A manufacturer must give a specific warning with a sufficient degree of specificity so as to cause a reasonable man to exercise for his own safety caution commensurate with the potential danger. It should be noted especially that

> the design of an adequate warning requires the combined talents of the engineers who determine where the dangers are and the text writers who convert test summaries into tintinabular messages. Together they shape the design events of the present to make available the best legal case in the future... An adequate warning...makes reasonable those dangerous defects or conditions in a product that the safety device cannot cure.[15]

Liability may also arise from the manner in which a product is packaged. A typical example involved a Coca-Cola bottle which fell out of a cardboard carton in a supermarket and exploded upon impact with the floor.[16] The court held that since it is

> common knowledge that bottles do not ordinarily fall out of properly made cartons which are not mishandled ... [and there was evidence that plaintiff had not mishandled the carton]... the jury could reasonably conclude that the carton was defective at the time of plaintiff's injury.[17]

Once the product consciously designed with adequate safety devices and warnings, enters the stream of commerce, the manufacturer's duty is by no means at an end. Aftermarket evaluation and retesting is a must, with a view toward discovery of any latent and hitherto unnoticed defects. In addition, and this point bears repetition, complete, detailed and accurate records must be kept of all pre and post market tests, inspections, surveys, etc., for the purpose of establishing any future defense.

In conclusion, the manufacturer should know that good manufacturing practices are his best defense to the threat of product liability suits. With that knowledge, and with the comfort that products liability insurance affords, he will be preparing tomorrow's defenses by indulging in good manufacturing practices today.

Footnotes

1. Santor v. A&M Karagheusian, Inc., 44 N.J. 52, 66-67 (1965).

2. Scanlon v. General Motors Corp., 65 N.J. 582, 591 (1974).

3. Restatement (Second) of Torts, §402 A (1965).

4. Moraca v. Ford Motor Co., 66 N.J. 454 (1975).

5. Id.; Jakubowski v. Minnesota Mining and Mfg. Corp., 42 N.J. 177 (1964).

6. Coccia, "Your Product is Defendable," 40 Ins. Counsel J. 137, 142 (1973) [hereinafter cited as Coccia].

7. McComish v. DeSoi, 42 N.J. 274, 282 (1964).

8. Huddell v. Levin, 395 F. Supp. 64 (D.N.J. 1975).

9. 391 F.2d 495 (8th Cir. 1968).

10. Clark v. Sears, Roebuck & Co., 254 So.2d 62 (La. Ct. App. 1971).

11. Bexiga v. Havir Mfr. Corp., 60 N.J. 402 (1972); Finnegan v. Havir Mfr. Corp., 60 N.J. 413 (1972).

12. Bexiga v. Havir Mfr. Corp., 60 N.J. 402, 410 (1972).

13. Coccia at 137.

14. Ruvolo v. U.S. Steel Corp, 139 N.J. Super. 578 (Law Div. 1976); See also Daleiden v. Carborundum Co., 438 F.2d 1017 (8th Cir. 1971); McCoy v. Wean United, Inc., 67 F.R.D. 495, (E.D. Tenn. 1975).

15. Coccia at 142.

16. Corbin v. Camden Coca-Cola Bottling Co., 60 N.J. 425 (1972).

17. Id. at 431.

CHAPTER 10

QUALITY ASSURANCE IN DIFFERENT INDUSTRIES

Reprinted from PRECISION METAL, June 1980

Quality control for investment casting waxes

Part I: Principles of thermal analysis
Modern analytical methods make it possible to achieve higher consistency in investment casting wax.

By MYRON KOENIG, technical director
M. Argueso & Co., Inc.

Thirty years ago, the manufacture of waxes for investment casting was a relatively simple matter. A limited number of raw materials were weighed out, loaded into a kettle, melted, blended together, filtered and packaged. Little attention was paid to quality control, and the investment caster used what was shipped.

Today, all facets of the investment casting industry have become more sophisticated. Wax manufacturing has progressed from an art to a science. Casting wax formulations now contain larger numbers of raw materials and are produced in much larger batch sizes. As a result, investment casting wax producers have had to devote more and more attention to quality control.

In the early years, only a few customers requested that any quality control tests be run. These were generally limited to melting point and ash tests. Over the years, a few other tests have been added, such as hardness, specific gravity and, where appropriate, filler content. As a result of this increased awareness of the need for quality control, the wax manufacturers of the Investment Casting Institute developed a investment casting waxes that was issued in November 1971.

Each batch of casting wax that we produce is tested by our laboratory prior to shipment. The tests that we run are:
- Softening point
- Hardness penetration
- Specific gravity
- Ash content

Although running these tests did much to improve the quality of our waxes, we found that the tests could not completely assure us that our products were as consistent as we desired. An occasional lot of marginal-quality material could be produced and shipped to a customer.

We therefore decided to investigate a variety of other analytical systems on the market. Many versions of chromatography, viscometers and thermal analyzers were examined. Each system gave some useful information, although not necessarily what we were looking for. We decided in mid-1971 to rent a DuPont Model 990 Thermal Analysis System on a trial basis. After a few months of debugging and experimentation, we purchased the unit for our laboratory.

Thermal Analysis (TA) in its broadest sense is the measurement of changes in physical or chemical properties of materials as a function of temperature. The visual observation of the melting of a compound in a capillary tube placed in a heated oil bath, so familiar to organic chemists, is a Thermal Analysis technique. Of course, modern instruments are much more complex and produce a much wider scope of pertinent information, but the fundamental principle is the same—the observation of a change in a material or system which is temperature-dependent.

Regardless of what temperature-dependent variable is being measured, TA instruments have certain elements in common. Figure 1 is a block diagram of a generalized TA system. The sample is placed in an environment, the temperature of which is regulated by a temperature programmer. Changes in the sample are monitored by an appropriate transducer or probe which produces an electrical output, the analog of the physical or chemical change. This output is amplified and applied to the readout device, usually a potentiometric recorder.

The temperature programmer must be capable of holding the temperature of the sample environment at some constant value. It must be able to vary the temperature as a known function of time (usually linear), the rate of which is selectable. Appropriate heaters and temperature-sensing feedback control elements are incorporated in the sample environment.

The transducer or probe is closely coupled to the sample and is specifically designed for the measurement being made. The amplification system is usually matched to the particular transducer; usually, high-gain, low-noise DC amplifiers are used.

The most popular output device is the x-y recorder, due to ease of interpretation and data storage. The dependent variable is plotted on the y-axis against the sample temperature on the x-axis. Thus, the variance with the temperature of the parameter under study is immediately obvious.

The four thermal analysis procedures in greatest use today are:
- Differential Thermal Analysis (DTA)
- Differential Scanning Calorimetry (DSC)
- Thermomechanical Analysis (TMA)
- Thermogravimetric Analysis (TGA).

We have found the first three to be useful for our purposes.

Differential Thermal Analysis (DTA)

DTA could be called the grandfather of modern thermal analysis methods. Differential Thermal Analysis involves recording the difference in temperature between a

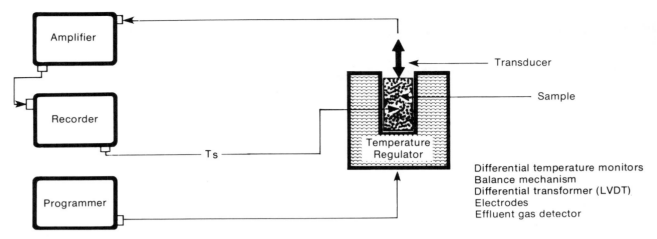

Fig. 1—Generalized thermal analysis system

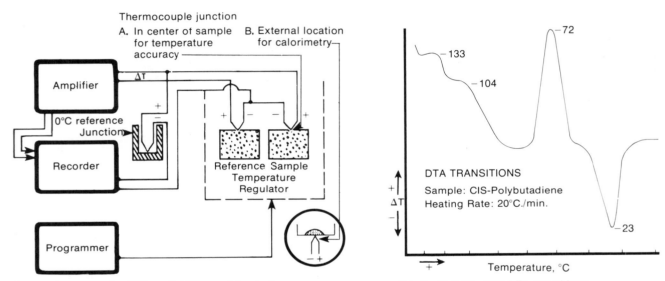

Fig. 2—Apparatus for Differential Thermal Analysis

Fig. 3—Plot showing DTA transitions

sample and an inert reference material as a function of the sample temperature. A typical DTA schematic is shown in figure 2.

The programmer, amplifier and recorder common to all thermal analysis systems are used as previously described. In DTA, the trsnsducers are a pair of thermocouples that are placed directly in the sample and the reference material. The reference material must be chosen so that no thermal changes are shown in the temperature region of interest. The thermocouples are connected in series so that they oppose one another. The resultant electromotive force (emf) of the pair is therefore proportional to the differential temperature of interest.

If the sample and reference are at identical temperatures, the resultant emf is zero, and the recorder shows a horizontal line. If the sample temperature is higher than the reference, the output emf is of a sign such that the recorder response is an upward curve in the DuPont system. When the sample temperature is lower than that of the reference, the sign of the emf reverses, and the recorder response is a downward curve.

The sample thermocouple serves a dual purpose—it is used as part of the thermocouple pair and also provides the sample temperature signal to drive the x-axis of the recorder.

Figure 3 is a DTA thermogram showing all types of transitions observed in a thermal analysis curve. The two transitions at −133C and −104C are second order transitions which indicate a slight change in heat absorption. The transitions at −72C and −23C are first order transitions which are distinct peaks. The transition at −72C is an exotherm indicating that the sample is releasing heat as the crystalline structure changes to a lower energy state. The transition at −23C shows the absorption of large amounts of heat as the sample melts.

Quality Assurance System For The Integrated Circuit

Masahide Takanashi
Manager, Semiconductor Quality Assurance Department
Toshiba Corporation

INTRODUCTION

The domain of electronics is covering a wide range of applications. The effect of product quality on user's economics has become the center of public attention, and the quality assurance of products has now gained a wide importance from various angles.

The International System of Authentication for electronic components is one of such certifications, and this sytem demands the satisfaction of reliability specifications in addition to the concept of quality assurance in the enterprise activities and the search for methods of its accomplishment.

Of the various electronic components comprising electronics, the semiconductor products, especially the integrated circuits demand ex-tremely high quality and reliability.

From the above standpoint, the present paper describes the process of quality assurance of integrated circuits employing new technology actively and the method of assuring reliability, these activities are the ones representing the company-wide quality assurance systems, and in particular, the quality assurance and reliability of C/MOS IC for communication industry, and active participation in National Quality Assessment Organization.

QUALITY ASSURANCE AND RELIABILITY ASSURANCE FOR THE INTERGRATED CIRCUIT

Systematization of Quality Assurance

Semiconductor products, particularly integrated circuits require short-term development, and this process often demands the adoption of new technologies. In the course of the developmental scquence, an overall quality assurance activities ranging from receiving orders of new products to trial production, large-scale production and services are needed in order to completely confirm their quality and reliability. For this purpose, it is necessary to elucidate the relationship and responsibilities between the related departments. These points are elucidated in Fig. 1 and the activities on quality assurance and reliability control are described in accordance with the flow of work.

Quality Assurance at the Development Stage of New Products

The main items of guidance activities at the development stage are summarized in (1) to (4), which must be practiced and considered in the quality and reliability review system described in the later part of this paper.

(1) Making market survey for the development of new products.

(2) Promoting the design and trial product development activities for new products.

(3) Establishing an organizational system that provides an opportunity to all engineers working in respective areas of survey, design, trial production and mass production to participate in Qaulity Assurance.

(4) Collection of data from various points and reliability review putting priority on the performance and the life, at the initial stage of mobilization.

Quality Assurance of Materials and parts

Components are particularly purchased ones, and the control and instruction are thoroughly exercised.

---

1. Komukai Toshiba-cho Saiwai-ku Kawasaki   JAPAN

(1) Realization of working plans (Systematization of Quality Assurance working program)

(2) Vendor control and instruction.

(3) Vendor auditing.

(4) Component acquisition system and receiving inspection.

Quality Assurance in Production Lines

Maintenance of quality and reliability in the production lines is based on the fundamental policy of self-reliance in the responsibilities involved and a priority is put on the in-process Q.C.

(1) Establishment of quality control standards (QCS) in the control systems of the production line

(2) Thorough corrective measures for process abnormalities.

(3) Follow up of process information such as the history of the process or the changes.

(4) Thorough equipment control and testing instruments control.

(5) Elevation of workmanship by small group activities such as ZD (Zero Defect) or Q.C. circles.

As to the inspection and control items, sufficient attention is paid to avoid major defects at the early stage of production by means of feed back of failure modes of IC products providing with the control of Al metalization by SEM at the wafer production stage or X-ray control at the molding stage. In addition, at the introduction of materials or new process the related production line is held stabilized by doing product tests and physical analyses in relation to materials adopted.

Quality Assurance of Production Lots

As for the integrated circuits, assuring of product lot is of great importance in addition to the quality guarantee in the production line, such as the In-process Q.C. stated above. All the products passing the production line are subjected to lot quality assurance tests and then stocked and kept ready for delivery. The quality assurance systems including the inspection of the production line are summarized in Fig.2.

Reliability Assurance

In addition to the lot quality assurance test, reliability tests or physical analyses stated above are performed at the recommended frequencies for the purpose of acquiring higher reliability or maintaining or confirming of the reliability. The results are feedbacked to the design and the production departments, and also utilized to the more complete reliability activities in the following items.

(1) Development of evaluation techniques for reliability.

(2) Development of techniques for failure analyses.

(3) Making of reliability control plans.

(4) Establishment of failure analysis method and study of failure mechanism.

(5) Collection of reliability data.

(6) Promotion of disciplines in reliability concept.

Service Activities

Service activities such as grasping the level of quality in the field or handling of complaints are important themes in the quality assurance activity. Particularly for handling complaints, procedures are clearly established so that rapid and accurate measures can be taken. The quality memorandum retrieval system into which information of discrepant goods handled are computerized provides appropriate follow ups. Other than complaints handling, the following are conducted as required.

(1) Follow up of quality levels both in and out the company. (Periodic meetings are scheduled)

(2) Analysis of the field data.

(3) Corrective actions by way of failure analyses (Fig. 3)

CONFIRMATION SYSTEM FOR QUALITY AND RELIABILITY

In obtaining products of high reliability, it is necessary that audits should be conducted on the basic design as well as production design and the process design including the stabilization of production lines. For this purpose, audits are conducted on several phases ranging from development to mass production, so that evaluation and verification of products can be made. The true purpose of audit is to make every engineer conscious of his reliability, to make clear the responsibility, and to incoporate reliability into the practical work. Based on the above, a system is formulated and put into practice comprising design reviews and reliability evaluation ranging from development stage of semiconductor product of systematized fashion to trial production and mass production stages. As is clear from Table 1, this system is divided into four stages, from development to standard product, and is evaluated and confirmed by the design review and confirmation system. The system aims at elucidating the responsibilities of each section concerned and harmonic operations of the parties related from the standpoint of technology, and quality and reliability are performed.

QUALITY ASSURANCE AND RELIABILITY OF C/MOS IC

Assurance of Design Quality of C/MOS IC

TOSHIBA's communication industry grade C/MOS IC employs a Clocked Inverter system generally described by C^2/MOS, and considerations are given to reliability design. Various investigations are made for design margin and high moisture resistance and so forth at the stages of both design and development and trial production for use in communication industries. For high moisture resistance the design of chips and packaging design are investigated for structures of frame and various kinds of resins.

Reliability of C/MOS IC

A number of reliability test data of C/MOS IC were accumulated from the accelerated reliability test at design phase and from periodical quality assurance inspection. Shows a part of their data collected. The acceleration factor for C/MOS IC can, also, be calculated to some extent. Table 2 indicates data for accelerated temperature in the accelerated operation life test and dilating curves are obtained as shown in Fig.4, and 0.6eV of activation energy can be estimated. These data offer ca. 20 Fits in reliability when the condition of the field is set as $25^\circ C$ and can maintain ca 100 Fits when the worst case is considered. In actual field data they represent several Fits to several tens of Fits presently, and these figures may be improved.

ACTIVE PARTICIPATION IN NATIONAL QUALITY ASSESSMENT SYSTEM

A plan for quality assessment system for electronic components is presently under way on international basis and in Japan too there exists a corresponding system, i.e., National Quality Assessment System for Electronic Components authorized by RCJ (Reliability Center for Electronic Components of Japan).

From its characteristics C/MOS IC has wide applications in measuring instruments and regulating instruments or others requiring high reliability. As our company has confidence in product quality which relates the increase in reliability of instruments to that of increase in components, our measuring and regulating instruments, for which high reliability oriented markets will offer chances for the more enhanced purchase, were subjected to the qualification of Quality assessment systems for electronic components of RCJ and were successfully registered as accepted products in terms of the integrated circuits and semiconductor products for the first time in Japan.

CONCLUSION

In order to maintain a favorable level of quality assurance for the integrated circuit, the necessity to ensure continuous and effective quality assurance, ranging from design phase to production phase, is strongly emphasizen as IC differs from discrete semiconductor device in complexity in design and production process and the speed of advance in process engineering. Standardization can be a possible solution out it offers a problem whether the standardization could cope with or not the rapidly changing pace of advances in IC technologies. From the standpoint of quality assurance, there remains much to be done. From now on we are searching for possible solutions and exercise product assurance more effectively, and thus hoping to provide a new stepstones for the development of the semiconductor industries in the world.

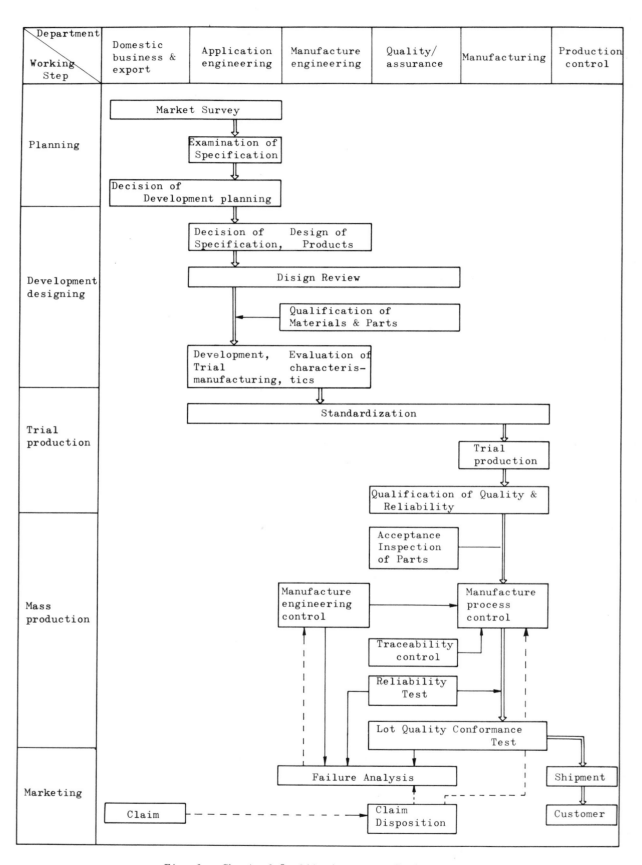

Fig. 1 Chart of Quality Assurance System

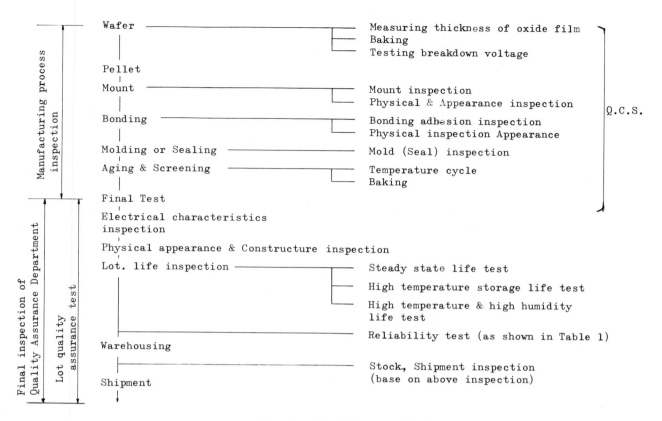

Fig. 2 Qualification Test System

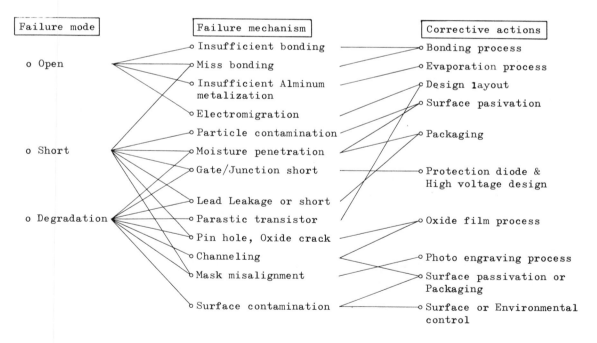

Fig. 3 Corrective Actions of Failure

Table 1 Control System in Product Development

Steps in development	Approval Test System	Participants
Survey, Review and Target setting for Quality and Reliability for successfull functioning of developed product	Design Review	Members from Development design, Applied technologies and Quality assurance
Considering economics and mass productibility, establish production process and Quality and Reliability	Design Approval Test (DAT)	Members from Applied technologies Production engineering, Technical engineering, and Quality assurance
Establish quality and reliability levels, process stabilization, QC methods	Quality Approval Test (QAT)	Members from Production engineering, Quality assurance, Fabrication group, and Production control
All of the control systems are established and standard products are produced.	When component material or production process is changed, DAT or QAT is exercised as priority indicates.	- do -

Table 2 Results of High Temperature Operating Life for C/MOS IC

Ambient temperature	Operating vaoltage	Total component hour	Failure Rate
85°C	V_{DD} = 16V	49.66 x 10^5	0.08%/1000H
100°C	V_{DD} = 16V	11.60 x 10^5	0.27
125°C	V_{DD} = 16V	48.68 x 10^5	0.48
150°C	V_{DD} = 16V	3.7 x 10^5	2.00

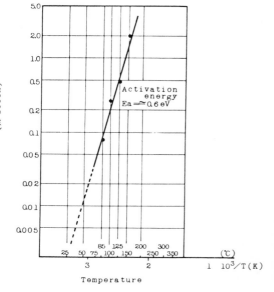

Fig. 4 Derating curve for C/MOS IC

Practical Experiences In Establishing Software Quality Assurance

by Edward I. Keezer
Office of the Project Manager
Army Tactical Data Systems
Fort Monmouth, New Jersey

Summary

The purpose of this paper is to furnish software managerial guidance to both government and private companies engaged in the acquisition and/or development of large scale computer programming systems. The paper discusses acquisition by the government of software and describes some of the specifications pertinent to the effective management of the contract. The paper also seeks to present the philosophy of "Participatory Management" practiced by our own organization. "Participatory Management" means active, frequent participation by the acquirer in furnishing guidance to the developer during in progress reviews, computer program testing, and internal and external controls and procedures. This approach is in contrast to an acquiring concept wherein the developer exercises his decision working perogative in a vacuum and shows a final product to the acquirer. Participation begins during the planning and formulation of early system design and culminates with the final acceptance of the software system. This philosophy is considered an effective way of ensuring that appropriate guidance is furnished to the developer and a realistic assessment of progress is made. Obviously those having the responsibility for a project should have a voice in the decision-making process. The meaning of the term is more fully developed during the remainder of the paper.

Introduction

This paper reflects the actual contractual experiences of various organizations associated with the Office of the Project Manager, Army Tactical Data Systems (ARTADS). Tactical Data Systems differ from commercial computer systems in their heavy emphasis on real-time and priority processing and their tailoring of the operating system to the needs of the contemplated tactical requirements. However, the managerial principles involved can be generalized to extra-governmental organizations as well.

ARTADS was established by the Department of the Army on 9 March 1971 with Brigadier General Albert B. Crawford, Jr., as Project Manager. The organization was formed to ensure the effective definition, development, acquisition, test, production, distribution and logistic and <u>software support</u> for assigned systems. Software is defined to include, as a minimum, all stored computer programs and associated documentation. The major active systems for which ARTADS has responsibility are TACFIRE, a tactical fire direction computer system for the Army's field artillery units, TOS, an on-line, real time information processing system designed to aid tactical commanders in the conduct of combat operations, and AN/TSQ-73, an air defense control and coordination system designed to coordinate actions of surface-to-air units against hostile air targets. The systems are in various stages of development, providing a cross-section of practical experiences.

Definition Phase

The first function Department of Defense charged ARTADS with responsibility for was effective initial definition of the system. By system, it is meant any self-sufficient combination of hardware, computer programs and personnel grouped together to perform an operational function or functions. The system specification developed is the definition phase. Thereafter, subsystem specifications are sometimes developed, if warranted, by the size and complexity of the original system. Finally, hardware and computer programming requirements are developed from the higher level specifications. The TOS program, for instance, contains two system specifications, three subsystem specifications, seventeen hardware specifications, and fifteen computer program specifications.

Contractual Requirements

Part I Specification

Care must be exercised on the part of any software acquirer in establishing the provisions of a contract, subcontract, or in-house development effort. Within the Government, the technical requirements document used to contract for a program or homogeneous grouping of programs is called a Part I Specification. The Part I is used to specify requirements peculiar to the design, development, test and qualification of the Computer Program. It establishes the initial design and forms the basis for the development of future test plans and test procedures. The specification could be equated to the rules of a football or basketball game, and the importance of the specification could be equated to the importance of both teams (acquirer and developer) following the same rules.

Other Contractual Provisions

The rules are also established through the use of Contract Data Requirements Lists (CDRLs) which are attached to the contract. These and their associated Data Item Descriptions (DIDs) provide for the format, content, and delivery schedule for deliverable reports and products, as well as for standards for design review and system software documentation. Failure to adequately specify the programming support software required for delivery on the TACFIRE contract resulted in the support software not being available in appropriate form for the eventual government takeover of that project or for future projects.

Initial Company Examination

Organization

After the contract has been awarded, ARTADS examines the organizational structure to ensure that the software manager is given a location within the company's organization sufficiently high to compete for resources and to present the software position on critical design decisions. He is not allowed to be subordinated to the hardware manager and left to cope with decisions made on hardware considerations alone. Worse yet, he is not left to design the software package in a vacuum with the expectation that somehow, miraculously, the system will interoperate at test time. Computer Systems Command, Fort Belvoir, Va., tasked with development of TOS software, had specific recommendations made to change its organizational structure to allow for increased control by the Project Office over the design activity.

Management Control

Next, the company is required to demonstrate the following:

1. Justification for required resources.

2. An orderly arrangement of tasks.

3. Internal Design, Test and Programming standards, controls

and procedures.

4. Detailed schedule of system software development allowing evaluation of activity times and critical paths.

5. Clearly identified responsibility for the various software development packages.

6. A method of accounting for and demonstrating the value of resources expended.

An additional explanation is required of item 3 above because of its significance in software development. Internal controls and procedures are those established by the contractor from previous programs and applied to the current development. ARTADS requires the potential contractor to submit his internal plan for controls and procedures as a part of his Technical/Management proposal in order to establish them early. Additional supplemental information is requested as a part of the Product Assurance and Configuration Management Plan. Before proceeding with the discussion on internal controls, a brief description of Configuration Management and its associated specifications used by the military for the management of software projects would be advisable.

Configuration Management

Configuration Management is a formal set of procedural concepts by which a uniform system of identification, control, and accounting is established and maintained for systems/equipment/computer programs and components thereof. Configuration Management military standards provide a means for standardization, and documentation, and management and control of computer program development. It is required by Army Regulation 30-37 to be applied to all Army systems, equipment and other designated materiel items throughout the life cycle of the materiel. Continued surveillance of the company's Configuration Management activities is not only used as a means for ensuring effective software development, but is also used by ARTADS as a principal gauge in measuring the quality of the developer's internal management. Most notable among the CM documents are the following:

1. MIL-STD-490, Specification Practices, which establishes the detailed format and content of specifications for computer programs and materiels. The MIL-STD has both a Computer Program Development and Product Specification. The development Specification describes the performance requirements necessary to design and verify the required computer program in terms of performance criteria. The Product Specification is the document representation of the Computer Program. It consists of flow charts and narrative that logically describe the design of the computer program, plus the listing or the series of instructions and data which constitute the Computer Program itself.

2. MIL-STD-483, Configuration Management Practices of Systems, Equipment, Munitions and Computer Programs, supplements and expands upon the documentation requirements of the previously mentioned standards. It also provides uniform procedures for preparing, formating, and processing changes to computer programs once a configuration is fixed.

3. MIL-R-83313, Reviews and Audits, established criteria and procedures for a series of technical reviews and audits. The reviews are conducted by the government with the participation of the contractor. Also, they are applicable to any acquirer reviewing any developer or for internal use only.

The in-depth nature of reviews called for in MIL-R-83313 can be illustrated by an example of the computer programming requirements of the System Design Review scheduled to be conducted early in the program. The review of computer programming requirements includes:

1. The computer programming techniques to be adapted for use in the system, e.g., on-line processing, off-line processing, parallel or multi-processing, multi-programming, time sharing, etc.

2. A gross description of the size and operating characteristics of all computer programs (e.g., application programs, maintenance/diagnostic programs, compilers, etc.) to include data base and other requirements.

3. A description of requirements for system exercising and identification of functional requirements (exercise configuration, conditions, missions, frequencies, functional simulation, recording, and analysis), and identification of major elements required to implement the exercising capability. (System exercising is that portion of the computer program dealing with the generation of simulated data and the analysis of that data).

4. Identification of all computer programs required throughout the system. Examples are: Operational programs; maintenance/diagnostic programs; test/debug programs; exercise and analysis programs; simulation programs; and compilers, assemblers and other required support programs.

5. Identification of all computer programming languages to be utilized in the system, and a description of how each language impacts the operation, maintenance and test areas. This MIL-R goes on to describe a similar series of reviews during the developmental period that culminate (considering other factors as well as audits and reviews) in acceptance of the programs.

Baseline

A term intrinsic to Configuration Management and to Internal Controls and Procedures is Baseline. Baseline establishment is an event during the programming process when, as a result of a review wherein the procuring agency approves or concurs with certain documentation, that documentation becomes an official portion of the baseline documentation of the system and hence subject to formal configuration management (change control) procedures. Internal controls and procedures apply at two levels:

1. The Design Requirement Baseline Level - in this case, the contractor's internal controls over the baseline ceases upon concurrence by the acquirer of the Part I (Development) Specification - at which point the requirements are baseline and subject to the MIL-STD-483 procedures.

2. The product configuration baseline established by the Part II (Product) Specification. The Part II Specification describes the exact configuration of the computer program as produced by the contractor. At this level the contractor's internal controls usually apply until the end of the developmental or acquisition period since the Part II Specifications, which establish the production configuration baseline, are often left open until the end of the formal qualification test period. Therefore, the following methods and procedures for maintaining and disseminating updated versions of the program must be in effect.

The provision and maintenance of a system of documentation (pros, flows, listings) which actually reflect the current state of the program and ensures that the program coding is traceable and properly reflects the approved design.

Methods, procedures and responsibilities for approving internally proposed changes to the product configuration baseline.

In-process reviews and audit procedures to insure compliance with established standards.

Defense Contract Administration Services Office

Now that the framework has been established, how does the Government ensure that the software products generated will conform to the MIL-STANDARDS and other applicable documents embodied in the contractual package? It is known that contractors deviate from their own internal procedures, to say nothing of those imposed upon by an outside agency. One of the methods used is to work in conjunction with the regional Defense Contract Administration Services Office to ensure that contractors adhere to prescribed documents as well as their own internal procedures. DCASO relies heavily on MIL-Q-9858, Quality Program Requirements, which requires the establishment of a quality program by the contractor to assure compliance with the requirements of the contract. However, the specification is completely hardware oriented and DCASO personnel lack expertise in computer software acquisition techniques. Therefore, ARTADS and DCASO have been working to rigorously enforce standards on such projects as:

1. Approval of the quality program established by the contractor to assure compliance with the requirements of the contract.

2. Auditing of the contractor's internal controls and procedures.

3. Auditing of internal configuration management procedures to ensure proper control of releasable programs and their associated documentation.

4. Ensuring that the coding represents the computer tape actually in operation and the supporting program documentation is a clear comprehensive reflection of the coding.

Test Requirements

Testing is generally conducted both by the programs within the design department and by an independent testing group. The responsibility of this independent test group is to review, evaluate, and test each program developed within the design departments. This testing group should discover errors and incompatibilities sufficiently early to prevent major redesign efforts.

The principal concern of the Independent Test Group is to ensure that all performance and design requirements, both implicitly and explicitly states, are verfied. Of course, testing of this magnitude is not achievable because of the enormous number of combinations which may be effected in any large scale software development package. Therefore, based on his knowledge of the system and software development, in general, the tester must select those areas where potential problems and incompatibilities exist and where sampling can be used to effectively ensure the reliability of related programs. Testing is conducted both at the individual program or module level and as the individual modules are combined into various assemblies. Careful consideration must be given to the level at which the individual module is to be tested in order to prevent duplication. Performance requirements must be quantitized where possible, and the sequencing of tests must be logical and realistic. Also, a sufficient, though cost effective, number and variety of input messages need to be scheduled in order to exercise the critical system limits, interface areas, timing factors, and storage allocations. There is also a requirement for a sufficient number and variety of illegal system inputs to be exercised in an attempt to make the system fail. Of course, these failure conditions should represent likely occurrences within the system and not merely unrestrained testing.

ARTADS carefully reviews all testing plans and specifications to be sure they incorporate the above requirements and sufficient time is allowed after each level of testing for the discovery and correction of problems. On-site surveillance is also effected in order to obtain an accurate evaluation of the company's internal testing and become aware of discrepancies or areas of potential disagreement in the documents (Test Plan, Procedure Reports). This allows us to input those technical and procedural positions that it is felt must be reflected in the final set of test documentation and concentrate resources on specific portions of the document which must be addressed at review time.

Now in addition to the lack of manpower resources, the software tester is traditionally deficient in the test equipment that is so readily available to his hardware counterpart. The tester should have in his baliwick test programs which ensure that the tested program complies to an established cording standard and that the test data is sufficient to thoroughly exercise the various statements and branches. Maximum efficiency in the use of core storage should be evaluated also. The TACFIRE contract demonstrated the difficulty of developing a complex software system in parallel with the development of programming support software. ARTADS is therefore attempting to provide engineering descriptions, functions, general specifications and tasking for software development and testing tools required for tactical system software support facilities.

Simulation Modeling

The use of simulation modeling is becoming increasingly popular on large-scale tactical computer development efforts. ARTADS has found the following advantages can accrue from the effective use of simulation techniques:

1. The ability to gain insight into the response of the entire system from a change in component design.

2. The ability to furnish information concerning the compatibility of proposed systems with existing systems.

3. The ability to determine the "saturation" points of various system components.

4. The ability to quantitatively evaluate the impact of design decisions. But often in practice, the simulation becomes large, unwieldy and unresponsive to the needs of the system under design. It also becomes extremely costly in terms of computer utilization and man hours. In extreme cases it even becomes an end unto itself, losing all resemblence to the actual system.

How does ARTADS avoid these pitfalls?

To begin with, experience has shown that approximately six months must be allowed to train even personnel knowledgeable in simulation techniques on a complex tactical computer development model. Therefore, ARTADS ensures that the simulation model is initiated as early as possible in the developmental effort. It also ensures that the contractor doesn't create a simulation package if an existing one can be adopted.

Also, specific plans for ensuring the active participation of the simulation team in the various planning and analysis design tasks are encouraged. Of prime importance is the coordination between the simulation effort and the design effort in order to keep the simulation current and ensure that results represent the actual system. Also, close surveillance is maintained over specification changes in order to keep the model current.

The confidence which design engineers have in the simulation package will, in large measure, determine its success. To foster this confidence, the company is encouraged to have the design engineers run simple simulation exercises, identify design alternatives requiring simulation, and help evaluate the results. The contractor is also encouraged to keep the design engineers informed of applicable simulation run results in their areas of interest.

Quality Assurance

This brings us to the last of the organizations to be discussed. ARTADS seeks to have the company establish a Quality Assurance Office whose principal responsibility in relation to the software development effort is to independently certify that the software products produced are of a quality nature.

The company's Quality Assurance Office is, in selected cases, required to make an independent review of all formal documentation to be submitted for Government approval. Now, independent review by QA does not mean that an aloof, hands off approach should be employed. Quite the contrary is true. It is only through involvement and participation in the emerging software design that the ability to evaluate the design's adequacy is established.

Of course, with limited resources, unlimited participation is impossible. Therefore, it is essential that the office select critical areas where problems are likely to merge as areas for prime consideration. This participation in no way precludes the reviewer from an independent assessment at the time of certification.

The eventual assessment may be done at two levels. First, to ensure that the software products are in accordance with good software quality control practices and in conformance to internal procedures, standards and specified contractual requirements. Secondly, they must be evaluated to ensure they conform to both implied and stated system requirements and specifications.

Some of the areas in which quality assurance must become intimately involved in are testing, standards development, programming standards, configuration management, review and monitoring of changes, design reviews, verification of performance requirements, and modeling and simulation.

Quality Assurance also has the responsibility for preventing organizational duplication and omission ("things just falling through the crack"). In order to effect coordination, quality assurance should assure that the managerial relationship, as well as procedures and methods for coordinating information flow between the various organizational units and tasks, are clearly defined. ARTADS attempts to confirm that the Quality Assurance Office performs the functions described above through personal contact, reviews and audits. On the TACFIRE contract, it is required that all submissions of Part II documentation must include a copy of the company's Quality Assurance certification of its completeness and accuracy.

Conclusion

It is noted that the concept of "Participatory Management" requires considerable planning about what is to be examined and how it is to be done in order to avoid interfering with the contractor or doing the job for him. However, the advantages to be accrued from effective participation include visibility and traceability during the developmental phase and an ability to modify, maintain, and understand the delivered product. It is also considered by ARTADS to be an effective means of ensuring timely delivery, avoiding cost overruns, and ultimately receiving a product that can do the job for which it was intended.

CHAPTER 11

QUALITY – A KEY ISSUE IN INTERNATIONAL COMPETITION

Product Quality: Theory and Practice in Multinational Approach

Fernando D. Negro
Fiat-Allis Construction Machinery, Inc.

IT IS NOWADAYS usually accepted that "quality" - one of the basic elements of success of a product together with "cost" - no longer means only "inspection" or "conformity" to design, but does involve all the functions, from product planning to design, to manufacturing, to service, to spare parts operation.

From an engineering standpoint, quality means to properly design and manufacture the products, in order to obtain the expected capability of implementing a function: this includes features, reliability, maintainability, durability, and safe operation. One of the simplest and clearest ways to represent these concepts is the following:

- the Product Effectiveness depends upon
  1) capability - (features, performance, productivity)
  2) availability - (% of possible working time).
- the availability depends upon
  - reliability
  - serviceability (includes maintainability and repairability).

Purchasing cost and durability, as they affect the operating cost, combine to give to the user an overall "quality" image of the product he is using.

ORIGINS OF MULTINATIONALS

Multinationals may have different origins, and this affects what we are going to examine in the area of product quality.

One origin is the gradual enlargement of a company, originally based in one country, to several new operations in other countries.

Another way of originating a multinational is by the merger of two or more companies, generally not too different in size, activity and

---

ABSTRACT

The control of product quality is now recognized to involve all functions of the manufacturing company, not only that of inspection.

Multinational companies have the problem of establishing a consistent quality assurance system with personnel having widely varying backgrounds and languages. Cultural and economic influences in different countries have produced different responses to the customer's demand for total product quality.

To successfully merge these diverse concepts into a practical corporate system, the multinational company must establish a central policy and program. A critical factor is the acceptance and support of the principles of the system by all levels of management.

maturity, formerly based and developed in different countries.

The first way of historical development of a multinational brings, almost spontaneously, the dissemination of one existing philosophy, one company policy, to all the growing worldwide operations.

The other way means, on the contrary, putting together different existing philosophies and policies, trying to obtain a profitable "merge" from what is more likely to be, at the beginning, a "conflict".

Many American companies have developed, during the last 30 years, into large multinationals of the first kind.

I belong to a company of the second kind, that can be defined a true multinational, as far as the origins are concerned. When the merger was made, industrial operations were existing in North America, in Europe, in Brazil.

I have the responsibility of product quality for this company after 26 years of engineering activity, the last 5 of them with increasing U.S. connections.

For these reasons, I have been exposed to ample opportunity (and to several shocks) of having to look at - and understand the differences of - various quality systems, in various geographical areas, extremely different in background, culture, mentality, historical and economical evolution.

EVOLUTION OF QUALITY SYSTEMS

As we are here basically an engineering oriented group, not a group of specialists in the quality field, it would be useful to attempt to briefly review the evolution of quality aspects and systems, in the most important industrial areas.

Up to World War II the quality function was nothing more than one of the several phases of the production cycle. The division of work, which changed the manufacturing system from artisanal to industrial, brought the typical "inspection" aspect of quality control. This lasted for several decades and is still hard to kill in some less sophisticated areas. The inspection activity pattern was obviously related to the size of the company and to the type of organization.

In the 50's, due to the increase in product sophistication, the higher level of technology and of people education, and a more demanding customer attitude, a double evolution took place:
a) as workers, individuals became less willing to accept the extreme division of functions: they were seeking more autonomy, more responsibility, more participation in the total process.
b) as customers, individuals became more conscious of their right to get from the products they buy the service they expect for the price they have paid. In other words, they were seeking the proper performance, reliability and durability relative to the cost.

We had, from that time on, different evolutions of the quality approach, according to the cultural and economical environment.

In the U.S., where the division of work was more widely and deeply established, the reaction brought the concept of a "total quality" system. The old inspection activity, and the subsequent quality control period, gave way rapidly to a "quality assurance" function, having the basic task of making certain that all departments, at all levels, were properly taking care of the final goal of customer satisfaction.

In Japan, where the industrial organization and the division of work had not yet attained such high levels as in the U.S., it appeared natural to utilize the willingness of individuals to participate in local and national effort for asserting the Japanese products (national pride), and the "quality circles" were promoted and developed. At the national level, educational programs were promoted to diffuse the knowledge of statistical disciplines, necessary to successfully approach the quality area.

In Europe, we may find various patterns, due to differences in local cultures and mentalities, and also to the different impact of North American influence. In less advanced areas we still find the "quality control" system, or even just the "inspection" department, reporting to a manufacturing manager. In larger, modern companies we can generally find the American system, or at least a first attempt to a local adaptation of it, with a quality assurance department reporting directly to the General Manager of the operation. Existing, but not significantly widespread, are some variations of the Japanese method, where all the employes, at all levels, are directly involved in the decisions concerning the production and quality organization.

It is probably well known to most of us that all modern quality programs and systems are emphasizing the prevention" factor, more than the "appraisal" (basically inspection) factor. This does involve more deeply in quality assurance all the functions which operate upstream of manufacturing: product planning, product engineering, manufacturing engineering, purchasing. It has been definitely proven that this approach, besides obtaining a better quality level, helps to improve productivity and costs.

Dr. W. E. Masing, at the recent Annual Conference held by ASQC in San Diego, made an interesting and clear comparison between American and German quality systems, and their effect on productivity, in the past (when labor cost in Europe was much lower) and at present (after the difference in labor cost has been substantially reduced).

When the main goals in American industry were high productivity and low cost, obtained through division of work and mass production, a definite quality problem was the result, he says. At the same time the German way to achieve high quality levels was through over-design and over-inspection. Obviously, this was acceptable while lower wages were there; when labor cost increased, the need of obtaining at the same time high quality and high productivity had to be faced. And new quality systems had to be defined and adapted.

PROBLEMS IN PRACTICE

All the above considerations belong more or less to "theory"; if by theory we mean the attempt to understand, describe and classify the principles of a human activity, or of a natural phenomenon.

The "practice" carries with it many more complications: the implementation of the basic principles is deeply affected by various factors that are rather difficult to predict and pre-evaluate. Consequently the results may largely vary from what was expected, and corrective steps are to be taken before achieving the objectives and goals that had been theoretically forecasted.

The experience has shown that more than one company, more than one system, have failed their first multinational attempts because of those practical factors not having been sufficiently and timely considered.

Different mentalities, different education, different cultural environment, different existing consolidated systems, require different phychological and tactical approaches. Different laws and regulations, different relationships with labor unions, will add further unknowns to the difficult task of finding the best way to promote and introduce a modern total quality system in a multinational company.

If we want to look somewhat deeper into the differences we have just tried to list, we will see that the American attitude is strongly oriented toward written and detailed policies, systems, procedures; toward precise and complete planning and budgeting every single function or activity, within the overall policy and planning of the company. The European attitude has always been quite different: the prevailing mentality, mainly in the neolatin Mediterranean areas, was that of "solving the problem when arising". In this kind of individually oriented attitude, every responsible person is supposed to face the situation and solve the problem with his own imagination, quick reaction, and invention: which is not so far from improvisation. Frankly, we must not overcriticize this kind of approach, because experience shows that it works, when the persons are really capable and know what they are doing (which is not always true). So we can say that while Americans are used and willing to have formal written documents telling them how to act, most Europeans seem to prefer, instinctivity, to be free to search out their own best way. Obviously these extreme positions - strictly "organization" and strictly "individualistic" - never occur in the reality; there are in between infinite nuances, and we may expect in the near future that the difference in attitude will be gradually reduced.

Another aspect that affects the practical implementation of a worldwide quality system in a multinational organization is the language barrier. It is not just a question of translation: different meanings of similar words, or corresponding meanings with different implications in local organizations, sometimes lack of corresponding words, may be cause of misunderstandings, or at least of difficulties in understanding. Anyone who has attempted to translate a procedure, trying to make it suitable and understandable for the organization to which it is addressed, knows too well how hard this task is.

We can conclude, from what is said above, that the major problems to overcome when introducing a new system in an articulated multinational enterprise are: 1) the unavoidable parochial attitude of each nationality, 2) the different mentalities of individuals and the different orientation of groups, 3) the language barriers.

Inversely, we must consider that the various existing ideas, mentalities, systems, if properly managed and utilized, can give a strong contribution to the enrichment and the validity of the entire system.

IMPLEMENTING THE QUALITY SYSTEM

At this point we should try to find an answer to the basic question that arises from the above discussion of problems and difficulties: "How can a quality system be properly implemented and managed in a multinational organization?"

Perhaps a complete and unique answer cannot be given, but certainly some general and specific suggestions can be outlined, according to both our own direct experience and similar experiences of other companies.

First, those who have the responsibility of designing the system and making it work worldwide, must be personally convinced that the basic doctrine they are developing transcends nationalities and applies to all entities, regardless of languages, customs, or attitudes.

Regardless of the origin of the company and the predominance of nationalities concurring to form it, the program must start from the top of the consolidated organization, down to the various functions, members, and local groups. This means that a Corporate Policy must be established; by the Chairman or the Chief Executive, clearly stating the will of having a system and the goals it has to achieve. Once this policy is issued, the further step is to make it understood and accepted by other high levels of the management: nothing can be done if they to not recognize the need of the program and the extreme importance of their support.

At this point at Corporate level a basic general program has to be prepared, and issued in the form of a manual containing the principles and guidelines for the implementation of the quality system by all operating units. This manual must be such as to be clearly understood and totally accepted by all the various entities of the entire organization. It must not be too theoretical, and at the same time not too detailed, in order to grant each unit or plant the necessary freedom in the implementation. But general goals and objectives, basic common standards have to be established, in order to avoid major discrepancies and uncertaintities. If we consider again the problems to be faced because of differences in mentalities, attitudes and languages, we easily realize that this is a very hard task. It is a task that requires a high degree of knowledge about all the aspects of the organization and the historical development of the company, if we are to avoid wrong moves that would require a long period of time to be corrected.

When these fundamentals - company policy, top management endorsement, corporate manual - are well established, it should be easier to carry on the further steps, typical of every modern quality system, such as:

- training local groups in total product quality concepts
- establish at all plants and operations a long range quality plan, and an annual plan
- issue at all plants a local quality manual, based upon the principles dictated by the corporate manual, but fitting the specific needs and possibilities
- assure the integration of all functional departments toward the attainment of quality goals
- obtain accurate and timely field reporting: this is the only way to take without delay the proper corrective actions, and establish and maintain a good image of the product
- design and implement an accurate quality cost system

- establish methods and criteria for the evaluation of system performance and results
- audit and measure progress, at least annually (possibly more frequently at the beginning).

I want once again to emphasize the fact that these steps can be easily accomplished <u>only</u> if the basic concepts have been properly defined, <u>only</u> if the system has been well asserted throughout the organization, <u>only</u> if care has been sufficiently taken for the different possible reactions of the various national entities.

IN CONCLUDING my reflections on some aspects of the quality system in multinational operations, I would add some suggestions, coming from the experience I went through:

- for the design and the implementation of a central quality system a professional must be hired
- in each national or local operation, the person in charge of the quality program must report directly to the person responsible for total local results
- the attention and the endorsement of top management (both central and local) on the quality system must be kept alive, and be constantly felt by the whole organization.

MANAGEMENT

Quality Technology – A Bridge to International Cooperation

by
Richard A. Freund
Sr. Staff Consultant for
Quality Assurance
Kodak Park Division
Eastman Kodak Company
Rochester, N.Y.

Sharing essential elements of quality technology will help all nations meet their quality goals in the 1980s

This article is based on a speech delivered at Institute of Quality Assurance Diamond Jubilee World Conference in London, England, Oct. 10 and 11, 1979.

The major challenges in the 1980s should lie in the continuing pressures for improved quality of products and services. This in an atmosphere of increased awareness of: the environment around us, shortages of both important raw materials and skilled craftsmen, and stronger competition as more nations become exporters of first-rate industrial products. In this latter regard, we only have to remember that a mere two decades ago Japan was not considered by the majority in the West to be an exporter of competitive consequence. Today we find South Korea, Taiwan, Singapore, and Hong Kong, among others, advancing along the same path that propelled Japan into the world of respected and highly saleable products.

Emphasis on health and safety aspects associated with the environment, including the disposition of waste material and hazardous substances, may ebb and flow because of economic concerns, but will not be reversed. This means that there will be need to replace many existing processes with others with which we are not as knowledgeable, and which will require very rapid, but also careful, development and optimization. Impact studies conducted to understand and account for the growing interdependence inherent in modern civilization will add further new dimensions of complexity.

Greater international competition in goods of relatively equivalent quality will cause further intensification of the drive to increase productivity. Emphasis will continue on automation of operations, testing methods, and data collection and analysis. New computer technology, both in software and hardware aspects, will make possible many advances which are merely in the formative stage today, or are currently limited only to some of the industry leaders. The lower cost of these powerful tools opens up many exciting possibilities. The role of the quality engineer should shift to greater managerial responsibility for achieving higher efficiency by reducing waste; by anticipating and preventing problems in production operations, and in customer usage; by obtaining better means of measurement and evaluation; and by becoming the master of that computer technology which will permit new advances in these areas. There is much talk today about increasing efficiency by giving the man at the bench a greater role through participative activities. In the same sense, the quality engineer must be given the opportunity to play a stronger role in management decision making with a greater awareness of financial implications and market requirements. The essential element of the coming decade will undoubtedly be a more systematic, organized approach to the economic control of quality. This includes recognition of the product and service requirements throughout their useful life, and of their total-life-cycle costs.

The role of management in the quality arena should also undergo a change in emphasis in the coming decade. It is management that is responsible for setting a clear and measurable quality policy. It is management that is responsible for providing the needed resources and for insisting on accountability.

There is an old and valid adage that "quality is everyone's business." Without everybody playing a role, and being concerned, quality is rarely achievable. On the other hand, it is also well known that what is everyone's business is no one's. Management is still responsible for seeing that its quality goals are implemented through efforts to achieve proper raw materials, employe attitudes, processes, and economic judgment. To do this task effectively calls for very specific line and staff assignments and the designation of key accountable individuals.

The environment of change facing us today calls for a reevaluation by management of its quality policies, and its determination to achieve those policies. Management must reevaluate its commitment to delegate sufficient authority and resources to the selected team of quality specialists so that appropriate advances can be made. A brief review of this environment of change may be appropriate at this point. It includes new product or service requirements; new operational concepts, new operational techniques or new basic materials; continuously changing national or international regulations and standards; and greater consumer sophistication and expectations.

Those charged with the task of dealing with these changes must view quality technology in a broad sense, encompassing development, implementation, and management of a quality system. I strongly feel that quality technology includes those management skills involved in assuring that all actions necessary to provide adequate confidence that a product or service will satisfy given needs are planned for and detailed.

Quality technology must also provide the operational techniques and activities that sustain a quality of products or services that meets these needs.

There are many arts and sciences included in the quality discipline. To manage the function effectively we must be alert to customer needs. There is a requirement to know the manufacturing or service capabilities. There also must be an understanding of the skills required, the tools and techniques available, as well as the need to recognize financial implications. Clearly, a key element is how to tie everything together — another way of expressing management.

Skepticism and Determination

Making the quality manager's role somewhat different from that of the development or production manager, is a sense of skepticism and determination to challenge what is planned or being carried out. This is done in order to anticipate and prevent whatever can go wrong. Prevention is almost a basic creed. Analysis is the fundamental tool of the trade. Since analysis plays such a key role in quality assurance, it is not surprising to find such specialties within this discipline as:

1) the science and application of measurement;
2) techniques for product or systems quality audits;
3) quality cost analysis in order to place emphasis where it should be;
4) statistics to deal with risks and uncertainties, and to effectively utilize the information collected, which is the basic commodity of the quality operation;
5) design analysis to evaluate the product or service in terms of satisfying the given needs, including those of the customer, and of society as a whole, and in terms of whether it will be possible to actually achieve the desired goals in normal production;
6) inspection techniques, including when and how to use acceptance sampling for incoming, in-process or finished products or materials;
7) reliability techniques to account for the effects of time or of complex assemblies;
8) maintainability, availability and many other related required abilities; and
9) human factors, and quality motivation since people continue to play vital roles in quality. Making it easier to do the correct job involves physical and motivational considerations.

The list could continue, and omissions are not intentional. They can all be summarized by emphasizing that customer needs must be translated into meaningful specifications; the design carefully reviewed, not only for function, but for service and maintenance aspects; facilities evaluated in terms of their capability of satisfying specifications; supplier quality monitored; product or service measured against the specifications; corrective action initiated and accomplished; and feedback recycled to create new or improved products as needed.

Perhaps of greater concern than these items are the organizational implications. Too often organizational questions seem to conceal the true functional essence of the quality operation. It shouldn't much matter where the quality function appears on the organizational chart, provided that the principles of independent assessment and the appropriate and timely feedback of information are achieved.

Obviously, all functions must be carried out in a coordinated manner, or there will be no system. However, flexibility in structure is vital to satisfy the many types of industry and operational arrangements. The important factor is that a system exists, is understood, is complete, and is effective. Styles of management differ too greatly to define the one best structure. Good communications are essential and performance is the best measure.

On an international scale, quality technology is advanced by the exchange of journals among many countries and through international conferences. Basically this communications effort is centered in the major quality assurance or quality control bodies around the world. The Institute of Quality Assurance (IQA) in Great Britain, the ASQC in the U. S., the Union of Japanese Scientists and Engineers (JUSE) in Japan and the European Organization for Quality Control (EOQC) in Europe are examples. The International Academy for Quality (IAQ), consisting of leading individual quality practitioners and educators from all parts of the world, was created out of the desire to extend this cooperation in an environment independent of any consideration of national boundaries. Individual teachers and practitioners have carried out the exchange practice for many years. Some elements of the United Nations, such as UNIDO, are occasionally involved. The recent Affiliated Society Agreement between IQA and ASQC represents the increasing formal effort to share our expertise on a world-wide scale.

Standards Community

On a more official and structured basis, international efforts are centered within the standards community which today is involved both with standards and product certification. Standards are basically a communications tool to define reasonable expectations. They serve national and international customers and suppliers. The International Organization for Standardization (ISO) and the International Electrotechnical Commission (IEC) are leading international structures in this area. The IEC is electronics oriented, while the ISO deals with most other products. The national standards bodies in each country serve as the source of expertise for these international groups, and the various national quality control societies should each work with the technical committees within, for example, the British Standards Institution (BSI), the American National Standards Institute (ANSI), the Japanese Standards Association (JSA) and so on.

The technical committees of these national bodies, as well as their ISO and IEC counterparts, develop product standards involving testing and evaluation procedures. Many of these groups could use professional advice on statistical quality control techniques as well as broad quality considerations. Today they often concentrate on specific evaluation criteria and methodology without relating to appropriate "fitness for use" criteria. Some product committees seem to use sampling plans that have repealed the laws of probability. Some seem to use unique approaches that have very limited validity. It doesn't make sense to standardize on unreality. This can only cause future trouble. Other committees recommend procedures that do not reflect many of the modern techniques which are almost second nature to the quality professional. For example, some of the sensory testing recommendations being considered within a food technical committee are truly mind boggling.

One area in the standardization process which needs consideration is establishing some technical review of generic approaches, either through a consultative function or through a stronger requirement of working with committees of discipline specialists.

There are other difficulties in the standardization process that are not easy to overcome. There is the fundamental question as to whether standards should detail all procedural steps or whether they should concentrate on functional requirements. This is a very important and highly controversial issue. Perhaps the efforts of the quality experts can help bridge some of these problems.

About the Author

Richard A. Freund is Past President and Chairman of the Board of ASQC. He was the first Chairman of the Publication Management Board, and received the Brumbaugh Award in both 1960 and 1962 for papers published in *Industrial Quality Control*.

A quality control practitioner since 1949, Freund is senior staff consultant for Quality Assurance at the Kodak Park Division of the Eastman Kodak Company.

Freund is active in quality control on the international scene, and is a member of the International Academy for Quality, an associate member of the European Organization for Quality Control, and Britain's Institute of Quality Assurance, the group to whom he delivered the talk upon which the foregoing article was based.

Some National And International Aspects of Quality

by Richard A. Freund
Eastman Kodak Company

The quality requirements for the 1980's will create new challenges for the quality professional. These will be discussed in terms of new approaches and modifications to the quality control efforts in this country and abroad, and in terms of national and international regulatory and standards efforts.

INTRODUCTION

It was a special pleasure for me, as well as an honor, to be asked by Corrinne Colombo to give the Youden Memorial Address today. Jack Youden always represented to me the ideal blend of practical and theoretical knowledge and application. I was fortunate to have the opportunity to meet frequently with him over about a 20-year span and to get to appreciate his personal warmth and technical grasp. This is why I feel that my topic today is in keeping with Jack's national and international awareness and activities. To put some of this in perspective, I'd like to talk a little bit about quality assurance activities in this country and around the world as they relate to the ASQC and the ASA.

FORECAST FOR THE 1980'S

The decade of the 1970's has witnessed remarkable advances in international exchange and cooperation to improve the quality of products and services throughout the world. However, this is only a prelude to the challenges of the 1980's. Pressure for the improvement of quality will continue to grow in an atmosphere of increased awareness of the environment around us, of shortages of important raw materials and of stronger competition arising from a general advancement in the quality of manufactured goods throughout the world.

The emphasis on the health and safety aspects, including the disposition of waste material and hazardous substances, may ebb and flow because of economic concerns, but this will not be reversed. This means that there will be need to replace many existing processes by others that we are not as familiar with, and which will need very fast, but careful, development and optimization. Impact studies for the purpose of understanding and accounting for the growing interdependence inherent in modern civilization, will add new dimensions of complexity.

Raw material and energy shortages should continue to rise during the foreseeable future. This will result in greater demands for efficient operations. Overdesign and some redundant features will become uneconomic to a higher degree, so that more effective use of quality assurance technologies will be required. The Japanese have demonstrated in recent years that methods some of the rest of us have been aware of as potentially useful tools, are, in fact, effective. In the next decade, many others will undoubtedly join this movement in the direction of paying greater attention to some of these "niceties", which now have been shown to work in practice as well as in theory.

Greater international competition in goods of relatively equivalent quality will cause further intensification of the drive to increase productivity. Automation of operations, equipment, testing methods and data collection and interpretation will increase. Quality engineering will play a key role in reducing waste by focusing even more on prevention and improvements rather than on appraisal. Metrology and methods of testing will call for more sophisticated effort, as will data analysis. New computer technology, both in the software and hardware aspects, will make possible many advances which are merely in the formative stage today, or are currently limited only to some of the industry leaders. The lower cost of these powerful tools opens up many exciting possibilities.

ROLE OF THE QUALITY ENGINEER

The role of the quality engineer should shift to greater managerial responsibility for achieving higher efficiency by reducing waste; by anticipating and preventing problems in production operations, and in customer usage; by obtaining better means of measurement and evaluation; and by becoming the master of that computer technology which will permit new advances in these areas. There is much talk today about increasing efficiency by giving the man at the bench a greater role through participative activities. In the same sense, the quality engineer must be given the opportunity to play a stronger role in management decision-making with a greater awareness of financial implications and market requirements. The essential element of the coming decade will undoubtedly be a more systematic, organized approach to the economic control of quality. This includes the recognition, and the activity which is based on such recognition, of the product and service requirements throughout their useful life, and of their total-life-cycle costs.

ROLE OF MANAGEMENT

The role of management in the quality arena should also see a change in emphasis in the coming decade. It is management which is responsible for setting a clear and measurable quality policy. It is management which is responsible for providing the needed resources and for insisting on accountability.

There is an old and valid adage that "quality is everyone's business". Without everybody playing a role, and being concerned, quality is rarely achievable. On the other hand, it is also well known that what is everyone's business, is no one's. Management is still responsible for seeing that its quality goals are implemented through efforts to achieve: proper raw materials, proper employee attitudes, proper processes and proper economic judgement. To do this task effectively calls for very specific line and staff assignments and the designation of key accountable individuals.

The environment of change facing us today calls for a reevaluation by management of its quality policies, and its determination to achieve these policies. Management must reevaluate its commitment to delegate sufficient authority and resources to the selected team of quality specialists so that appropriate advances can be made. A brief review of this environment of change may be appropriate at this point. It includes: new product or service requirements; new operational concepts, new operational techniques or new basic materials; continuously changing national or international regulations and standards; and greater consumer sophistication and expectations.

Those charged with the task of dealing with these changes must view quality technology in a broad sense which encompasses the development, implementation and management of a quality system. I strongly feel that quality technology includes those management skills involved in the assurance that all actions necessary to satisfy given needs, are planned for and detailed. Quality technology must also provide the operational techniques and activities which sustain a quality of products or services that meets these needs.

QUALITY TECHNOLOGY

There are many arts and sciences included in the quality discipline. To manage the function effectively one must be alert to the customer needs. There is a requirement to know the manufacturing or service capabilities. There also must be an understanding of the skills required, the tools and techniques available, as well as the need to be cognizant of the financial implications. Clearly, a key element is how to tie everything together, which is another way of expressing management.

What makes the quality manager's role somewhat different from that of the development of production manager, is that sense of skepticism and determination to challenge what is planned, or being carried out. This is done in order to anticipate and to prevent whatever can go wrong. Prevention is almost a basic creed. Analysis is the fundamental tool of the trade. Since analysis plays such a key role in quality assurance, it is not surprising to find such specialties within this discipline as:

1. the science and application of measurement;

2. techniques for product or systems quality audits;

3. quality cost analysis in order to place emphasis where it should be;

4. statistics to deal with risks and uncertainties, and to effectively utilize the information collected, which is the basic commodity of the quality operation;

5. design analysis to evaluate the product or service in terms of satisfying the given needs, including those of the customer, and of society as a whole, and in terms of whether it will be possible to actually achieve the desired goals in normal production;

6. inspection techniques, including when and how to use acceptance sampling for incoming, in-process or finished products or materials;

7. reliability techniques to account for the effects of time or of complex assemblies;

8. maintainability, availability and many other related required abilities;

9. human factors, and quality motivation since people continue to play vital roles in quality. Making it easier to do the correct job involves physical and motivational considerations.

The list could continue, and omissions are not intentional. They can all be summarized by emphasizing that the customer needs must be translated into meaningful specifications; the design carefully reviewed, not only for function, but for service and maintenance aspects; facilities evaluated in terms of their capability of satisfying specifications; supplier quality monitored; product or service measured against the specifications; corrective action initiated and accomplished; and feedback recycled to create new or improved products as needed.

ORGANIZATIONAL IMPLICATIONS

Perhaps of greater concern than discussing these items are the organizational implications. Too often organizational questions seem to conceal the true functional essence of the quality operation. It shouldn't much matter where the quality function appears on the organizational chart, provided that the principles of independent assessment and the appropriate and timely feedback of information are achieved.

Obviously, all the functions must be carried out in a coordinated manner, or there will be no system. However, flexibility in structure is vital to satisfy the many types of industry and operational arrangements. The important factor is that a system exists, is understood, is complete and is effective. Styles of management differ too greatly to define the one best structure. Good communications are essential, and performance is the best measure.

INTERNATIONAL ASPECTS

I was fortunate last year to attend the International Conference on Quality Control in Tokyo and a meeting of the International Academy for Quality in Kyoto. A short time ago I also attended a meeting of the European Organization for Quality Control in Budapest, and the Diamond Jubilee Conference of the Institute of Quality Assurance in London. All had a very strong, common theme. Quality is an international concern. It is a basic element in world trade, and such trade is fundamental to peaceful coexistence. Trade is affected by economic and quality barriers. Treaties involving certification of a product on a common basis are being developed rapidly, and these involve many statistical and other quality control considerations. Product liability actions and consumer protection regulations and legislation are growing almost exponentially. I was interested to see the enormous emphasis on consumer quality now being pursued within the Socialist Bloc. A new 5-year program in the Soviet Union is placing great attention on this area even to the extent of becoming extremely active in international standards and quality control bodies and conferences. But before I go into detail on standards and regulations, let me review some observations about quality control in Japan.

QUALITY IN JAPAN

Before World War II, Japan was noted around the world for producing poor quality merchandise. In general, if a product such as a toy bore the label "Made in Japan", it was expected to be junk and short-lived. Even the Japanese tended to buy industrial equipment abroad, rather than risk reliability problems. Following its defeat in 1945, it became obvious that Japan could only survive as an industrial nation, and then only if it could change its quality reputation. One approach was the formation in 1946 of JUSE, the Union of Japanese Scientists and Engineers, for the purposes of promoting interest in the advancement of science and technology, and of contributing to the development of industry through the exchange of information. JUSE is an independent foundation authorized by the government, but funded through industry contributions, dues and fees. An early move by the leaders of JUSE and the American occupation authorities was to ask Dr. W. Edwards Deming of the U.S. Bureau of the Budget to develop a series of lectures for top management in Japan on the importance of statistical methods as a management tool. He also was asked to develop several series of lectures for quality control engineers and statisticians in industry. Those of you who know Ed Deming can be sure that he emphasized his basic statistical principles for management in which he:

1. distinguishes between sources of variation a worker is responsible for, and those which management must react to;

2. the consideration of uniformity as a quality characteristic in its own right as contrasted to occasional superior and inferior occurrences;

3. dealing with system faults as contrasted to special cause situations;

4. using statistical tools to minimize over-adjustment or of doing too little;

5. control of the measurement process;

6. collection of appropriate data for analysis at the production level as well as at the management level;

7. statistical fundamentals in consumer-research and testing of product as actually used.

In short, he emphasized product performance as an interaction between the product, the user, and how the product is used, installed and maintained and how effective are the instructions, the training and the service. The Japanese have named their highest quality honors for Dr. Deming in tribute in his key role in reversing their quality image.

Naturally there were other essential elements in this successful transformation. I heard the Japanese introduce Dr. Deming as the father of their quality control and Dr. Juran as the "wet-nurse". Joe Juran is another management consultant who keeps his eye on the target. He concentrated on many of the non-statistical management approaches, but never ignored the value of sound analysis in focusing attention where it should be focused. He introduced a broader orientation of quality control as an integral part of the management function. In Japan, this newer orientation became known as Total Quality Control, but it is different in certain areas of emphasis from the Total Quality Control introduced earlier in this country by Dr. Feigenbaum. Dr. Feigenbaum's well-known book was also influential in this process. Of course, the guiding light in JUSE and in Japan, is Dr. Kaoru Ishikawa, who is the effective embodiment of modern Japanese quality assurance developments and educational efforts.

I found it interesting to learn that after top management first became committed to Deming's approach, and later to Juran's, there was a step-by-step program to bring everybody on board. Middle management, scientists and production engineers attended classes and discussions. In early 1960, radio programs were broadcast aimed at the production worker, and shortly thereafter classes were even given via television.

A later, but highly successful, movement was the Quality Control Circle, developed and nurtured by Dr. Ishikawa. This is basically a structured participation program. Groups of about 10 or so workers from a particular production line would volunteer to stay after work to undertake studies and to offer suggestions as to how to improve quality. They would choose a chairman, define a meaningful problem they saw and develop a solution to the problem. To aid in this problem-solving venture, they were taught some basic statistical and other analytical approaches. They learned about histograms, control charts, Pareto distributions and the Ishikawa "cause and effect diagrams", and in some cases went on to experiment design, analysis of variance and regression analysis.

These quality control circles were registered through the JUSE which sponsored various competitions based on judging the effectiveness of the results of a circle's project. Prizes were awarded within a plant, within a region, and eventually within the country. Some of you have seen the winners of such national competitions tour the United States as part of the recognition they received for outstanding successes. Today, there are about 90,000 circles registered by JUSE, and probably an equal number of unregistered ones sponsored by other groups. This represents more than 900,000 participants, which as Joe Juran has indicated, has been estimated to have produced the equivalent of some additional 10,000 engineer-years work over the past decade. This is not simply a motivational program as we first viewed it, but a real participative activity in which the basis of statistical thinking penetrated deeply into the operator level.

CONTRASTS WITH THE WEST

Not only operators, but engineers became knowledgeable about these fundamentals. It is interesting to note the massive education programs employed in many of the major Japanese industries. We offer much, if not all, of the same opportunities within the western world, but again there is a difference in culture and in approach. A design engineer in most major Japanese companies is required to take reliability courses during working hours. We tend to offer these courses on a voluntary basis either at work or as a tuition-paid night program. I am not attempting to suggest what we should do, but only to show another way East and West differ in attitude and culture.

Another contrast is our overall approach to quality control or quality assurance. In many Western companies, fairly sizable quality departments are responsible for inspection and quality engineering. Statistical engineering is an off-line service. In Japan there seems to be a tendency to place the responsiblity for these functions within manufacturing. However, a quality assurance nucleus guides and reviews the emphasis on analysis and prevention. Another oversimplified way of describing the difference is that in the West we tend to automate our way around problems, while the Japanese use their data to direct their preventative efforts through rigid data analysis. Detractors of this approach say it leads to always looking backwards. Supporters say that experience is an excellent guide. One suggestion that I would like to offer is that the quality control specialist and the statistician in this country should start to pay attention to the literature from Japan and other nations. Enough of it is in English to make this practical. There are a wealth of application as well as theory articles that could benefit us, even though some modifications undoubtedly will be required to fit our circumstances.

I have talked quite a bit about Japan because the quality improvement there has been so dramatic. Please don't get the idea that within Japan this has been a universal achievement. It hasn't been. Many small Japanese companies have managed to withstand this type of enlightenment, and struggle along. However, vertical integration of vendors into a large company's quality system has negated such problems for the key export industries. Much of the gain by Japanese industry represents government support through taxation benefits for capital improvements, and from encouragement to enter markets having higher than average pay-out possibilities in terms of export trade. The antagonistic confrontations between labor and management in the West is echoed here in the relationship between government and business. In Japan this does not exist, and there is cooperation to increase the Japanese market. However, there is no question that a management-supported quality effort based on both statistical and total quality management approaches has made real progress. I am merely pointing out that clearly it is not the whole answer.

The next phase of the Japanese program is to concentrate on market research. The me-too-ism period is to be replaced by innovation, with attention to the needs of the local consumers around the world. We are already beginning to see signs of this activity. Here is another area in which the United States must be active. One reason for this new thrust is that Japan now has a few additional problems to face. Their standard of living has climbed so rapidly that labor is no longer cheap. Its neighbors in Taiwan, Singapore, Hong Kong and Korea have learned their techniques and are competing with lower labor costs and improving quality. Japan has become the quality model for a large portion of the world. In fact, the Ishikawa Medal in Taiwan, plays the same role as the Deming Medal in Japan. One reflection of the Japanese role as a quality model is the sizable quality circle activity within Korea and Brazil, for example.

STANDARDS AND CERTIFICATIONS

But it's time now to get back to a different international aspect. This is in the area of standards and product certification. Standards are basically a communications tool to define reasonable expectations. They serve both national and international customers and suppliers. The International Organization for Standardization (ISO) and the International Electrotechnical Commission (IEC) are leading international structures in this area. The IEC is electronics oriented, while the ISO deals with most other products. The national standards bodies in each country serve as the source of expertise for these international groups, and the various national quality control societies should each work with the technical committees within the various standards organizations. In the United States these standards writers are represented internationally by the American National Standards Institute (ANSI).

ISO and IEC technical committees, as well as their American counterparts, develop product standards involving testing and evaluation procedures. Many of these groups could use professional advice on statistical quality control procedures, for example, as well as on evaluation criteria and methodology. Some product committees seem to use sampling plans that have repealed the laws of probability. It doesn't make sense to standardize on unreality. This can only cause future trouble. Other committees recommend procedures that do not reflect many of the modern techniques which are almost second nature to the quality professional.

Of course, there are difficulties in the standardization process. Fundamental questions arise such as whether standards should detail all procedural steps or whether they should concentrate on functional requirements. This is a very important and highly controversial issue. Perhaps the efforts of the quality experts can help bridge some of these problems. In the past, most Americans have taken the attitude that standardization is not for them. It restricts ingenuity. It limits improvements. Today a good share of the third world countries have included compliance to international standards in their legal systems.

There is work underway to form a new ISO Quality Systems Committee, and a counterpart within the IEC. These will need the active participation of knowledgeable people. As international certification of products expands, as it is doing within the framework of the ISO-CERTICO (the Certification Committee of ISO) and the IEC-Q (the IEC Quality Certification system) programs, there will be a continuing need for assistance. The new GATT (General Agreement of Tariffs and Trade) Treaty has a large section dealing with product certification, and requires a national inspectorate body. We are already beginning to see federal regulations concerning the accreditation of testing laboratories appear in the FEDERAL REGISTER.

These clearly involve both quality systems and statistical quality control implications. In the statistical quality control area, many feel that sampling plans are something best tailored for each situation and that a collection such as ISO 2859 or the U.S. MIL-STD-105D was a simplistic system intended for the non-thinking inspector. Among the things I have learned in my standards activities is why there still is a need for a universal standard of this type, and also how flexible, as well as complex, it really is, when used as intended.

If you really want to quarrel about optimization of sampling plans, consider the deep doubt about the validity of any sampling evidenced only a few years back at the hearings before the Consumer Product Safety Commission. Individual plans may be more efficient from our point of view, but excessive individualism does not make the public feel confident about our intentions. We live in a very mistrusting environment today. In my opinion, what is needed is a reasonably flexible set of standards with logical ground-rules for selecting the appropriate approach. This is what the ASQC Standards Committee, the ANSI Committee Z-1 on Quality Assurance, ASTM Committee E-11, and ISO/TC69 are all considering. Needless to say, there are many viewpoints offered.

In the national picture, I have just mentioned ASQC and ANSI, who along with ASA and ASTM are becoming more active in quality standards. In this era of regulation and legislation, this has become a necessity. Like many of you, I used to feel that standards were dull and usually unnecessary, and often not very intelligent. Today, where standards don't exist, government agencies will hastily write their own. Many are reasonably handled, but not all. When we prepare standards in the voluntary concensus system, they are widely circulated to those who volunteer to review them and who agree to offer constructive suggestions. The late Wisconsin Congressman, Representative William Steiger, once told a meeting I attended that industry was then complaining about the OSHA legislation, but that when he asked for comments during the development of the bill, he received no help from industry despite his pleas. If we want sensible statistics and quality systems in standards and regulations, we are going to have to work with the standards writers to head off wrong approaches and to develop sound ones. My experience has taught me that standards work is anything but dull. Working with people like the late Harold Dodge, John Mandel, Acheson Duncan, Norm Johnson, Stu Hunter and Hugh Hamaker, to name only a few, is always stimulating and educational.

PREPARING FOR THE FUTURE

Perhaps all of this can be summarized and tied together by asking what the role of the statistician or quality assurance expert is today. Most of us employed by industry and government, as well as a majority of those in the academic community, have found out that the person who calls for assistance in solving a problem usually makes the actual decisions about what action is to be implemented. Those of us who consult on design approaches, serve to analyze data, and to suggest conclusions, or evaluate quality or quality systems, play an important role in leading to these decisions. The key to this role is our credibility. To be effective, we must clearly demonstrate the solution and not add to the existing confusion through poor assumptions or incomplete definition of the problem. The Japanese approach has been to make more production and development people aware of how to work with and interpret data, and to let them do so.

I see the role of the industrial statistician and the quality control engineer evolving into developing the many untapped but fertile fields for analysis, such as the revitalized approach to graphical techniques; the new concepts and breakthrough approaches; the training and guidance of a broad engineering, scientific and production community. We are seeing some slight signs of this already, but in my opinion we have not begun in earnest the type of new activity that will let us fully utilize the country's potential. Jack Youden had the vision of making his knowledge available in a form that others could use by themselves for the most part. Perhaps that is the crux of the issue. Many feel that a powerful tool in the hands of an amateur or semi-professional may be more dangerous than no tool at all. This feeling holds in medicine, in education, in psychology and other fields as well as in statistical analysis.

Maybe management must take the first step in the breakthrough cycle by encouraging a reasonable level of education and training coupled with a reward system that recognizes effective use of consulting groups. By effective use, I mean avoidance of the pure service function syndrome, and emphasis on the review and evaluation of the procedures and results of the user by a professional, in addition to counseling, in the true sense of quality assurance.

While quality assurance depends on the sound analysis of data, the analytical mind of the professional in this field must deal with the entire range of considerations, and not merely on the methodology. The concept of "fitness for use" must be paramount. The customer, whether external or internal, must be the prime concern in the fullest sense of quality which encompasses economic considerations. Quality is everyone's business, but it still requires a strong and clearly identified central focus and full-time dedication.

CHAPTER 12

FUTURE TRENDS IN QUALITY ASSURANCE

Programmable controllers, minicomputers, and microcomputers in manufacturing

J. L. Miller

The need for new techniques in manufacturing has become increasingly critical during the past several years. Factors that contribute to this need are 1) increasing complexity of products, 2) severe pressures of domestic and foreign competition, 3) increasing consumer demand for higher quality, 4) influence of governmental and other agencies for greater safety and reliability, and 5) inability to increase prices to keep pace with increasing production costs. The electronic and other businesses of RCA demand manufacturing development to keep pace with product development so that RCA can retain its competitive position.

James L. Miller, Director, Manufacturing Systems and Technology, Corporate Staff Manufacturing, Cherry Hill, N.J. graduated from Iowa State University in 1948 and joined RCA as an engineer in the Component Parts Department of the Tube Division. During that time he designed printed circuit i.f. amplifiers and rf tuners. In 1953 he transferred to the Home Instruments Division, Color TV, designing i.f. amplifiers, color demodulators and monochrome video circuits. In 1960 Mr. Miller moved to the Computer Division where he worked on high-speed tunnel-diode circuits and later became a manager of peripheral equipment design responsible for card readers and optical character readers. In 1966 he joined the International Licensing organization as the Managing Director of RCA Engineering Laboratories in Tokyo, Japan, and, in 1969, joined Corporate Staff Manufacturing as a manager and later as the Director of Manufacturing Systems and Technology.

Reprint RE-21-3-2
Final manuscript received June 23, 1975.

COMPUTERS have been available for many years to provide the speed and capabilities which made many complex and, at times, impossible tasks practical for the first time. Most of these early systems were quite large and expensive and were used in such applications as financial control, scientific and engineering calculations, communications, *etc*. Because of their complexities and cost, only limited use was seen in manufacturing. Introduction of programmable controllers, minicomputers, and microprocessors during the past few years is making it possible for the first time to develop practical manufacturing systems for automatic test, data collection, and process control in manufacturing.

Programmable controllers

Although not as flexible and powerful as a minicomputer or microprocessor, programmable controllers can replace complex relay banks with logic switching that conforms to Boolean expressions. Programmable controllers offer several advantages. They are more flexible and easily changed than static logic or relays. As the complexity of control increases, they offer a cost advantage at about 40 to 100 relays over standard switching logic. Reliability and life of solid-state programmable controllers generally far exceed the expectancy of relay logic. The ability to easily change logic statements

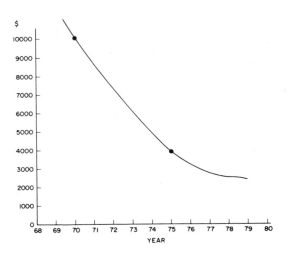

Fig. 1 — Minicomputer prices.

gives the controller greater flexibility when compared with the need to modify wiring in a relay panel. This is especially useful when designing and testing a new system. Since most suppliers provide straightforward methods to program the controller, it can be easily used by personnel familiar with relay switching logic without the need for computer programming skills. Early users were in the automotive and machine tool industries but installations are now being implemented in process control of chemical, petro-chemical, food processing, paper and pulp industries. In some applications, programmable controllers are used as primary control devices with minicomputer backup for the purpose of loading programs, monitoring process conditions and providing more complete control.

Minicomputers

In addition to programmable controllers, minicomputers are being used more extensively in manufacturing applications. One contributing factor has been the drastic reduction in minicomputer prices during the past five years as seen in Fig. 1. In addition, both hardware and software capabilities of minicomputers have been expanding during this same period. Many of the functions previously provided only in large computer systems are now available in these devices. Large core memory at decreasing prices along with such features as hardware multiply/divide, memory protect, automatic restart, etc., have become common. In software, such things as real time operating systems, file management, and process monitor-direct digital control packages are available from one or more of the minicomputer suppliers.[1] In addition, high level languages such as Fortran, COBOL, BASIC, etc., are available on most equipment.

Multiplexing capability permits sensing process conditions, inputting of test results, or entering data manually from many points around a manufacturing floor. The ability to add memory and peripheral devices, as needed, makes it possible to tailor systems to meet specific requirements. This added capability over programmable controllers brings with it the requirement that programming personnel are usually required for effective use of the minicomputer. The most important ingredient of a successful installation is careful planning. A thorough study of the application and its requirements should be completed before consideration of a vendor and the specific software and hardware approach to be used.

Microprocessors

The most recent device to make its appearance in the control scene is the microprocessor. This device has most of the Central Processing Unit (CPU) capabilities of the minicomputer on one or two chips. Added chips are provided for memory, input/output control, etc.

Microprocessors, in general, are lower cost with smaller memories and lower speeds than minicomputers. Early applications appear to be as controllers for testers, process control, and machine tools. Again, with more capability than the programmable controller, they require a higher technical competence to use and maintain. The use of programmable controllers, minicomputers and microprocessors in manufacturing varies over a wide gamut including machine tool control, testing, facility monitoring, process control, automatic assembly, data collection, injection molding, and warehousing.

Applications

New applications in manufacturing and product control are being introduced at a rapidly accelerating rate. It is expected that this trend will continue in the foreseeable future to meet the pressures of cost, reliability, and safety. Two general areas and one specific example from the thousands of specific applications will illustrate the breadth of use.

Automatic warehousing

Automatic warehousing has exploited the minicomputer extensively and is offered by several companies, such as SI Handling, FMC, and GE. These systems include control of trucks for placement and retrieval of product, and range in complexity from operator-assisted to completely automated warehousing.

Injection mold control

Several packaged systems are available using minicomputers from such companies as IBM and Harrel, Inc. These provide monitoring and/or control of cavity pressure, melt temperature, dwell time, etc.

A system developed by the Rochester product division of General Motors Corporation uses General Automation's SPC-16 minicomputers for carburetor adjustment and testing with a group of 31 minicomputers reporting to an IBM System 7 for supervisory control. The System 7 acts as a communication concentrator providing information to a 370/145 at the management level. The System 7s have disc storage in case the 370 goes down.[2]

RCA has been active in the application of

minicomputers to manufacturing during the past few years in the areas of automatic test, data collection, and process control.

Automatic test

Some of the earliest computer applications in manufacturing have been for automatic test. The need for thorough and sophisticated testing of components, subassemblies, and complete systems by Government and Commercial Systems has made it necessary to develop rather extensive capabilities in this area. The Government Communications and Automated Systems Division in Burlington, Mass. has extensive experience in this area, and additional capabilities have been developed in the various divisions to meet specific needs.

The test center concept is being used effectively in Camden. A complex of three RCA 1600 computers is used for data input to control testers for logic systems, analog video equipment, and communication equipment; the complex also includes diagnostic capability. Additionally, a General Automation SPC-16 is used for system testing and a DEC PDP-8 for magnetic headwheel balancing. Most modules produced in the Camden Plant are tested at this center.

The semiconductor industry has been another early user of automatic test. The need for high speed, accurate testing of devices has become increasingly important as complexity has progressed from discrete transistors to integrated circuits (ICs) to medium scale integration (MSI) and finally to large scale integration (LSI). One of the first in-house developed test systems was the Computer Controlled Automatic Test equipment (CCAT) for probing of wafers.[3] This system uses an RCA 1600 computer and special equipment controller with three extenders, each of which controls three wafer probe stations. Data from these test systems is logged and taken to an off-line computer for analysis including histograms and wafer mapping.

Hewlett-Packard equipment has been used extensively for testing of power devices. This testing started with the development of a system using the Hewlett-Packard 2114 minicomputer controlling three test stations to qualify components. Since that time, continuous development has been in progress.[4]

One of the first linear testers developed in RCA was the Automatic Dynamic Universal Linear Tester. (ADULT), which uses General Automation's SPC-12 computer as the controller.[5] This system is aimed at specific circuit configurations with separate test stands for each IC type. The SPC-12 provides switching of stimulus and measurement equipment and pass or fail decisions.

General Automation SPC-16 minicomputers are also being used to measure glass faceplates and funnels for color picture tubes. Initially started as a means to measure faceplate contour, it has since been extended to include funnel and other measurements.

Systomation in-circuit testers are being used successfully for testing component values after assembly and dip soldering into printed circuit modules at Consumer Electronics. These machines measure the value of most components to within $\pm 1\%$. Since the average test time using this equipment for a normal size board is in the order of 5 to 10 seconds, it provides a very powerful tool for testing those devices which do not require adjustment. Since the system provides printout of components not meeting specification, the complicated and sometimes inaccurate troubleshooting steps can be eliminated. Additional work is progressing to use computer control for testing adjustable modules and subassemblies.

The use of an Intel 4004 microprocessor for the testing of horizontal output transformers has been developed by the CE division. Tests performed include control of an indexing table and measurement of inductance ratio, phase and high voltage breakdown with the advantages of reduced test time and increased test accuracy. In addition, the microprocessor keeps an account of good and reject transformers. In all of these test systems, the inherent possibility exists to gather product information and enter it into defect reporting or manufacturing data systems. This provides manufacturing and engineering with a tool to identify repetitive product faults and implement corrective action to further improve product performance and reliability. In most areas where automatic test equipment is being used, plans are progressing to further improve utilization through the use of resulting data output.

Data collection

An early system developed for collecting information from the manufacturing floor is the Defect Reporting System (DRS) developed in Consumer Electronics. This application provides input terminals located at troubleshoot positions on the manufacturing floor tied directly into an RCA 1600 computer. The data is compiled in report format and printout is provided by teletypewriters located at strategic points. These early systems were aimed at collecting defect information as it occurred and presenting it to management in a form making rapid corrective action possible.

Another defect reporting approach being used is Automatic Defect Diagnosis System (ADDS), with manual defect logs used by troubleshooters and inspectors as the source of information. This data is keypunched and entered into a 70/45 computer for analysis. The system is quite flexible and can be configured by each user to fit his own needs. Another feature permits the user to select that portion of data in which he is interested such as specific line number, model, type of defects, *etc.*, by the use of delimiters when requesting reports. These delimiters permit examination of the total data base or specific portions of interest. The ADDS system is very flexible in its application and, therefore, has been or is being used in such places as Consumer Electronics in Taiwan, Government and Commercial Systems, Consumer Electronics and Appliances in Canada, and Electronic Components in Canada. Applications include analysis of field defect information as well as in-plant use.

Another development providing not only defect reporting capability, but also product tracking through the manufacturing process is the Manufacturing Data System (MDS). This uses a General Automation SPC-16 computer and input terminals similar to those used by the Defect Reporting System (DRS). Greater flexibility is provided as the user can configure the entire system by filling in tables which define data validation routines, input terminal locations, data meaning and data names to be printed in subsequent reports. The user can request reports at any time, giving product status, product flow, in-process inventory, and defect analysis. A unique feature of this system is the X-Y report which permits the user to list any combination of two items, one on the X axis and the other on the Y axis, with the body of the report showing coincidence of the number of counts or quantities summed for these two parameters.[6]

Data collection systems for specific divisional needs have also been developed. Video terminals as input devices are being used by Solid State Division in conjunction with an IBM System 7 processor. The terminals are used in a community mode in which data is entered by clerical personnel as product enters and leaves a functional area such as final test. In addition, data concerning the results of tests can be entered for later analysis. As in most computer controlled data collection systems, the user has the capability of on-line examination of data in a real-time manner.

Although not completely manufacturing oriented, Electronic Components developed an Order Control System (OCS) using a General Automation SPC-16 computer which provides warehousing control throughout the country at various locations. Customer orders are entered by clerical personnel through video terminals at each location and an examination of the data base as to product availability and inventory status is provided. The system printout permits warehouse personnel to select product and also provides billing information to include with the packaging.

Process control

Process control is many times thought of as the complete closed-loop control of a total manufacturing plant such as those systems which have been installed in the petro-chemical industries. Although this type of complete control is usually impractical in most of our manufacturing processes at RCA, it is many times possible to identify small process steps which will yield very effectively to tighter and better control. For some years, certain components have been automatically assembled into printed circuit boards at Consumer Electronics. This approach uses the United Shoe Machine equipment in a continuous line process, assembling one component at each station. Although very effective for high volume runs with few changes, line changeover due to mixtures of products is a time consuming and expensive procedure. Recently, both United Shoe Machine and Universal Instruments Corporation have made available stand-alone component insertion equipment which utilizes X-Y tables to position the workpiece under the insertion head. The insertion head then inserts each component in turn until all insertable components for that board have been put into place. The control mechanism for this device is a General Automation SPC-12 computer which is fed from a paper tape prepared off-line and is used to instruct the machine in the proper sequence of X, Y and Z positions. The computer also adjusts the insertion head to accommodate the size of the component to be inserted. Sequencing machines are used ahead of the insertion equipment to place components on taped reels in the proper sequence so that parts will go into their proper location.

Process control systems using the DEC PDP-8 computer have been developed to adjust to the thickness of latex applied to tufted carpeting and the speed of carpet through drying ovens. Microwave transmitters and receivers are used to sense the amount of moisture in the latex material, since the microwave signals are attenuated as a function of the amount of moisture in the carpet. With this system measuring the moisture content at the output of the drying oven, it is also possible to adjust carpet speed automatically for various carpet styles since the oven temperature is considered a long-term constant. In addition, gas usage monitoring has been added which has proven to be effective in adjusting oven efficiency.

Process control has been developed for the production of photomultiplier tubes at the Electro-Optics and Devices activity in Lancaster. Initially using a PDP-9 and later upgrading to a PDP-11, this system controls the rate of deposition of photomultiplier material on the anodes of the tube and makes specific tests of the product as it is being processed.[7]

In the production of integrated circuits, many critical steps need to be monitored and controlled. Included are the furnace diffusion steps which require tight control of the variables as well as verification that all process steps have been satisfactorily completed. To make these more positive and reliable, many furnace suppliers are developing control systems based on micro- and mini-computers. Solid State Division has recently ordered furnace control systems, one using a microprocessor with card reader input to control insertion and withdrawal rates, setpoints on temperatures, and gases. The second system will include a minicomputer which feeds recipes to the microprocessor and provides a certain amount of supervisory and reporting capability.[8]

Extreme care and careful planning must be exercised when applying process control. People with intimate knowledge of the process and the control system who can effectively communicate and work toward a common goal should be available. In most cases, process variables are interactive; therefore, a period of process monitoring is usually needed to establish correlations to verify intuitive relations established during manual operation. Special attention must also be given to equipment reliability and backup procedures in process control.

Conclusion

RCA is actively involved in many areas of manufacturing technology. The techniques necessary to provide acceptable yield when producing large scale integrated circuits, color picture tubes to extremely tight tolerances, competitively priced tv sets with high reliability, and other quality products have been under development for some time. Only a few of the projects involved with automatic test, data collection, and process control have been outlined in this paper.

Since manufacturing is a major contributor to product cost, performance, reliability, and customer acceptance, increasing attention must be centered on this part of our business.

References

1. Lazar, M.H.; "Merit of direct digital control, *this issue.*
2. *Production* (April 1974)
3. Lambert, H.R.; "Computer controlled automatic test system, *RCA Engineer*, Vol. 21, No. 2 (Aug - Sept 1975) p. 20
4. Guerin, W.R.; Lyden, T.B.; Martin, M.J.; McLaughlin, G.; and Ruskey, T.J.; "DART: Diffusion Area Rearrangement Task — open heart surgery on a living factory," *this issue.*
5. Fowler, V. and Roden, M.; "Adult II test system" *RCA Engineer*, Vol. 21, No. 2 (Aug - Sept 1975) p. 30.
6. Carrell, R.M.; "The computer—its use in managing manufacturing operations," *this issue.*
7. Hallermeier, S.; "Computer controlled processing of phototubes," *to be published.*
8. Dillon, L.; Van Woeart, W.; and Shambelan, R.; "Product control system for semiconductor wafer processing," *this issue.*

A look at QC circles

by Robert T Amsden*
Associate Professor of Management
University of Dayton
Dayton, OH

and Davida M Amsden
Author/Lecturer
Huber Heights, OH

As a nation, we in the United States have prided ourselves for being rugged individuals, going it alone in the face of great odds. But more and more, as the decade of the 70s drew to a close, we came to see that there are other approaches: specifically, the task-team. For example, educators have found that team teaching is an effective way of communicating knowledge to children. In labs, corporate boardrooms, and think tanks, men and women have begun to pool their ideas, talents, and insights to solve problems.

During the last five or six years, nearly a hundred companies in the US have started using teams to solve work-related problems at the shop-floor level. These teams are known as *QC Circles*. What is significant here is that, up until now, most applications of the team approach in industry have been at higher levels of a company. Indeed things have come full turn because there are staff, engineering and management QC circles that use the same format, process and tools as hourly circles.

Basically the QC circle is a problem-solving group of about 4 to 12 members. Whenever the membership consists of hourly paid employees, the people come from the same work area, whether it be from one section of the assembly line, the machine shop or another department. In most cases the foreman acts as circle leader, although there are situations where an hourly employee leads.

The circles meet on a regular basis, usually for an hour a week with pay. Some firms have them meet during working hours; others schedule meetings outside regular hours. Another ingredient most companies stress is *voluntary participation:* no one is forced to join, though circle members or management certainly may *encourage* involvement.

The focal point of circle activities is the identification, solution and implementation of the solutions to work-related problems. The problems solved may increase quality levels, reduce waste and rework or increase efficiency. It is important to understand that circles concentrate on technical problems; they *do not* address issues that properly belong to management or organized labor.

A circle goes to work

At this point it will be helpful if we present a hypothetical circle project. This will illustrate both the problem-solving process and the tools the circle members use to work through their project. In **Figure 1** we have sketched the problem-solving process.

Dr Robert Amsden is a member of the ASQC and Chairman of the Steering Committee on QC circles. Together with Dr Davida Amsden, he co-edited the ASQC publication, QC Circles: Applications, Tools, and Theory.

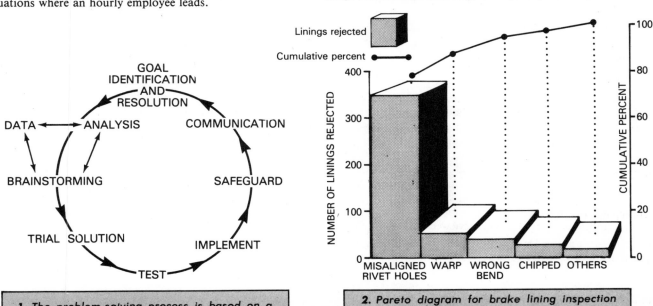

1. The problem-solving process is based on a closed-loop feedback system.

2. Pareto diagram for brake lining inspection from March 1 to 15. QC checked 1800 pieces and rejected 471.

The brake-lining department of an automotive parts manufacturing firm has a new QC circle. The Breaker One Nine QC circle consists of five males, three females and the foreman, who is their leader. Having just completed 10 hr of training in how circles solve problems, they are eager to begin their first project.

During training, the circle members looked at all the problems they could identify in their area. One problem stood out as being of long duration: the excessive rate of defective brake linings. Indeed, one of the quality-control inspectors had told them during a training session that this problem was one that the QC department had wanted to correct for some time. Thus the circle members selected the brake lining as their first project.

Because they were not certain of the relative importance of the various types of brake-lining defects, they needed some data from the QC department. They asked QC to send an inspector to the circle's next meeting to bring the appropriate inspection data. In between the circle's weekly meetings, the foreman contacted the chief QC inspector and talked with him about the circle's purpose and needs, asking for his assistance and cooperation.

The inspector came to the next circle meeting. The foreman opened the session by welcoming the inspector and reminding the circle members that this was not a time to "go get QC." Rather it was an opportunity to learn from them and to assist them by solving one of their problems. He stressed that, since the circle members were the ones who produced the brake linings, perhaps they could help to improve the product.

The inspector then reviewed the inspection data with the circle. They decided that a Pareto analysis, one of their problem-solving tools, would help identify the major types of defects. The leader then asked for three volunteers to develop a Pareto diagram based on the data from the QC department. The volunteers said that they could use breaks, lunch time, machine shutdowns and after hours to do the analyses, so that it would be ready for the next meeting.

Isolating the problem

The following week the circle reviewed the Pareto analyses of quality problems. The three members who had prepared the diagrams pointed out that the one for the period March 1 to 15, shown in **Figure 2,** was representative of the last 12 months of inspection records.

The Pareto diagram revealed that one item, misaligned rivet holes, caused 75 percent of the defects. Other types of defects were minor in comparison. One circle member commented, "that's a tough one to solve. We've had that problem for the last eight years." The rejoinder, to which he agreed, was that if the circle were to make a real impact on quality it would have to do so where the opportunities lay—and that was in rivet-hole alignment.

The foreman said that the inspection data very accurately portrayed a real problem, that the personnel who attached the brake linings to the metal shoes had a great deal of difficulty in assembling many of the shoes because the rivets would not fit through the misaligned holes.

Once the circle came to a consensus that the project was to be the rivet-hole alignment, the foreman led them in *brainstorming* for causes of misalignment. First he assigned one member to record all brainstorming ideas on the flip chart. As leader he then went around the circle asking each member to give one possible reason for the problem. He cautioned them not to evaluate any of the ideas at this time, even the seemingly far-out, wild ones. Occasionally a member passed as his turn came up. The brainstorming continued around the circle for nearly 12 min. Three volunteers agreed to put the results of the brainstorm session into a *cause-and-effect diagram* and have it ready for the circle's next meeting.

When the circle next met, the members began with a review of the diagram, **Figure 3.** The three who had prepared it explained that the problem, rivet holes out of specification, appeared in the box on the right side of the diagram. Major cause areas—man, machine, method and material—were shown as the main "bones" of the diagram. Sub causes were listed where appropriate, so that the diagram was actually an organization of the brainstorm ideas. The foreman asked the circle to spend another 10 min to brainstorm any further ideas to add to the diagram.

The QC circle then studied the cause-and-effect diagram

A look at QC circles ...continued

"A QC circle concept is...management's doing something with the workers, not to them."

in great detail. They decided to go back to the process itself and examine the oven cure times to see if they were too short. The circle also studied the boring machines to see whether setup might be the true cause. The leader divided the group into two subcircles, one for each study. At the next full-circle meeting, one group reported that examination of cure times was a dead end. The time that brake linings were in the cure oven was very carefully fixed at a constant value; temperature of cure was also constant, as indicated by automatic recording charts.

The other subcircle, however, excitedly told that their first combined sample of 100 linings taken from both boring machines showed 49 percent outside of specifications, mostly on the low side. A further check indicated that all of their sample of 50 from Machine B were within specifications as shown on their frequency histogram, **Figure 4.** Apparently this machine could hold to the specifications. However, the other machine had 60 percent out of spec, some on both sides.

In the ensuing discussion, circle members pointed out that Machine C probably had not been set up correctly, but, even if it had, the excessive variation in the process would have caused a large percentage of output to be out of spec. They also reviewed a check sheet which indicated that most of the misaligned holes occurred at the ends of the linings.

They checked these linings by using a single metal shoe and fitting each lining onto it by hand insertion of rivets. The QC circle felt that it was time to invite an engineer to their next meeting to go over their findings with him and to seek his advice on how to proceed.

At the subsequent meeting, an engineer agreed with the circle that most likely the problem of rivet-hole misalignment resulted from a combination of drill setup errors and excessive variation during the boring process. He pointed out that Machine C was long overdue for a complete overhaul. He also advised them that, if management decided to replace or overhaul the machine, it would take some period of time. An interim solution was necessary.

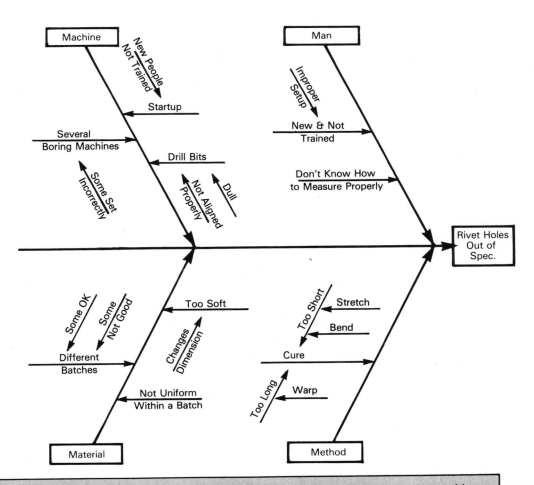

3. *The cause-and-effect diagram or fishbone chart helps find the major cause of a quality problem.*

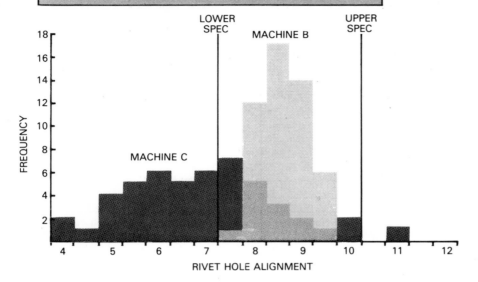

4. Frequency histograms of two machines sampled April 10. Machine B makes all parts in spec, Machine C does not.

Circle recommendations

The QC circle decided to recommend the following to management:
• The long-term solution—overhaul Machine C.
• The interim solution—drill only with Machine B. The capacity of Machine B would enable them to meet daily production requirements, provided that a few hours a week were done on the second shift. The circle then prepared a presentation for management to obtain approval for their recommendation.

In the management presentation, each member of the circle had a part by presenting a portion of the project or having prepared some of the charts and diagrams. The production superintendent, plant engineer and maintenance manager all attended.

Within the week, management approved the recommendation. At the circle's next regular meeting, the foreman told them that management was very pleased with their problem-solving and documentation, and had accepted their recommendation, with the interim solution effective immediately.

The circle decided to monitor the results during the interim phase to discover the effects of removing the troublesome Machine C. They depicted the first results in a graph, **Figure 5**, which indicated an appreciable improvement. Even so, the circle wanted to explore other avenues that might further reduce the defect rate.

Breakthrough

This "case study" clearly shows how the QC circle carries through the complete problem-solving process:

• Identification of problem areas.
• Selection of problem by the circle (suggestions may come from outside).
• Solution of the problem by the scientific method.
• Implementation of the solution (with permission of management where required) by the circle.
• Maintenance of the solution.

This example points up a major aspect of the circle process: specifically that the circles have been engaged in *breakthrough*. In **Figure 6** we have a chart that pictures the difference between control and breakthrough. For years a certain level of defects had been tolerated as standard operating procedure. As long as the defect rate stays within the accepted limits, the process is said to be in *control*.

Occasionally the process goes out, and it is the job of production with the assistance of others to bring it back into control as soon as possible. By contrast, the Breaker One Nines changed the *control limits* and reduced the percentage of defective workpieces. This happened over a period of many weeks because it was not something that could be accomplished quickly.

A system where management listens

Through the years, several concepts have been developed as problem-solving methodologies. However, we feel that there are very basic, fundamental ideas embodied in the QC circle that are not combined in other concepts. No other concept integrates *all* of the particular elements of the QC circle.

The Kepner-Tregoe problem-solving method provides an excellent means of solving problems, but it is designed for

A look at QC circles
...continued

use by one individual. The QC circle is, first of all, a group process that solves problems. Integral to this process is brainstorming, a technique that necessitates group participation. Second, the circle meets on a regularly scheduled basis. K-T does not require this.

Swedish job-reform activities, in particular the *production group*, were designed to enlarge the jobs. The intent seems to have been one of motivating hourly employees by reforming the job and wage system. By contrast, the thrust of the QC circle action is solving work-related problems through the scientific, data-based principles that are used by the employees themselves.

With the circle, work and its organization continues *as usual*, except that the circle members continually scrutinize the technical aspects of the job and thus are constantly developing improvements. Workers who are part of the QC circle are actively engaged, not just in doing their jobs, but also impacting their own work situation for the good. For many it is the first time they have been able to affect their work life in a positive way.

Another way in which the QC circle concept is unique is that it is an instance of management's doing something **with** the workers, not **to** them. So many motivational programs, while coming from all kinds of good intentions, are *perceived* as top-down and exploitive.

One of the most distinguishing features of QC circles is training. Typically it begins at top management levels with a one-day seminar introducing the concept. Then facilitators, circle leaders and various personnel such as staff, engineers and union representatives go through an intensive three- to five-day training period. This covers the circle concept, group dynamics, problem-solving tools used by the circles, and instruction in how to teach these problem-solving tools to the employees.

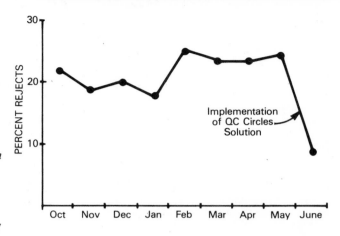

5. Graph shows workpiece reject rate before and after corrective measures.

As the task-team is becoming the setup for scientists, teachers and managers to cooperatively do their work, so the QC circle has become the avenue for hourly employees to work together to solve their problems. Increasingly, management is recognizing that the hourly people in the factories and plants are the greatest untapped resource for quality and productivity improvement. The circle concept encourages workers to become active, cooperative participants in the future of *their* companies. Further, the circle gives the workers the problem-solving tools to do their jobs right and to affect the environment where they work. There is hope that they can change things for the better. ■

Selected Bibliography:
Amsden, Robert and Amsden, Davida (ed), *QC Circles: Applications, Tools, and Theory*, American Society for Quality Control, Milwaukee, 1976.

Juran, Joseph, *Managerial Breakthrough*, McGraw-Hill Book Co, New York, 1964.

Kepner, Charles H and Tregoe, Benjamin B, *The Rational Manager*, McGraw-Hill Book Co, New York, 1965.

Kondo, Yoshio, "Commonlines and differences in the Swedish job reform and Japanese QC circle activities," 20th EOQC Conference, Copenhagen, Denmark, June, 1976.

Transactions, of the First and Second International Conferences of the International Association of Quality Circles, Feb 1979 and Feb 1980, Cupertino, CA.

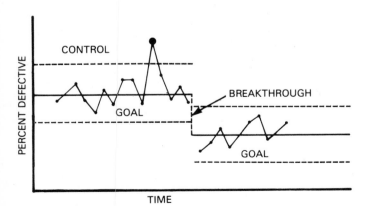

6. Chart shows quality going out of control, then settling down to new goal after breakthrough.

Reprinted from PRODUCTION ENGINEER, November 1979 issue "Quality Circles—Japan's Way To Better Quality"

Quality Circles – Japan's way to better quality

Involve your workers in your quality problems – that's the message the Japanese want to export to British industry.

QUALITY control in western industrial countries suffers from managers who still treat workers as so much plant and equipment.

This is the message which emerged from a recent London conference on Japanese 'quality control circles'—the method which has aided Japan's industrial growth over the past 15 years and is now being imported into British, American and European factories.

The gap between the Japanese and other industrial nations in 'quality consciousness' was outlined at the conference by Professor Kaoru Ishikawa of Tokyo University, who, as Director of the Union of Japanese Scientists and Engineers (JUSE) and chairman of the Japanese Society for Quality Control has been instrumental in introducing these 'quality circles'.

Quality circles were introduced in 1962 and have since been promoted by the 500 members of the JUSE, which runs extensive courses in quality control techniques for supervisory staff throughout Japanese industry. Quality circles are basically small groups of workers normally from the same workshop, who meet regularly to discuss quality control problems and their solutions, usually under the guidance of the workshop foreman.

All workers in each factory belong to a group—each consisting of between 8-10 members. At the end of 1974 there were thought to be one million such groups operating in Japan.

Group members are encouraged to discuss quality problems and work defects and are introduced to modern quality analysis techniques, such as pre-control charts, pareto analysis and cause and effect diagrams. Where possible methods which can be used to solve workshop problems are presented to the workers by practical applications and audio visual techniques so that beginners are not involved in complex arithmetical calculations. These are backed by repeated group discussion, exercises and 'training games.'

Elite

Professor Ishikawa stressed the importance of the foremen in the circle system. They form the nucleus of quality control activities, leading the groups and contributing to operators' workshop training.

Japanese foremen tend to be given a greater degree of responsibility than their Western counterparts and take charge of quality, cost, quantity, and the maintenance of equipment, jigs, tools and instruments, says Professor Ishikawa, reflecting, in his view the basic philosophical differences between Japanese and Western industry.

"Problems have arisen in the West because of the adherence to the 'Taylor' system of management. This means that engineers draw up work standards and machine operators work to those standards. There is no room for innovation on the part of the worker or for any account to be taken of his opinions. People are treated in the same way as machines."

But in Japan and in most of the Western countries today, the situation has changed, he added. The Taylor system could not be applied to Japanese workers since they were well educated and have an advanced knowledge of culture and technology, he said.

"If the system were to be imposed on them they would lose interest in their work, absenteeism would increase and we would see a rise in labour turnover."

Later in the conference, David Hutchins, MIProdE—a technical consultant with Executant Ltd, which organised the conference—reiterated the criticism of the Taylor system:

"It relies on the creation of a management elite which passes instructions on to a subservient workforce. Under such regimes almost all responsibility is removed from the foreman and passed to functional specialists such as quality engineers, work study engineers, production engineers and inspectors. Its advantage is that large quantities of cheap goods can be produced as compared with our pre-war craft-based production methods."

But, he added, it has many disadvantages. "The foreman has been divested of his responsibilities and these have been placed with specialist management groups who do not identify with the workforce and often make decisions which are not wholly acceptable to them."

Hence, leadership of a type attractive to management had been destroyed. If suitable leadership were not provided by management, the workforce would produce its own leader, he added.

Mr Hutchins pointed out that during the past decade industry had attempted to introduce job enrichment, zero defect campaigns and various forms of worker participation.

"All have produced disappointing results when compared to the recent success of the Japanese. The reason is clear. They are all based upon the Taylor management principle and no motivation programme can work properly in such a dehumanised environment."

In the UK the Quality Circle system has recently been adopted by Rolls-Royce, whose quality engineering manager Jim Rooney outlined the company's programme.

"In the past Rolls-Royce was dependent on the skills of its tradesmen in transplanting design requirements into the final product. But the ever increasing need for dilution of skills due to increased technology made a new approach essential," he explained.

The phase of passing from high-skill technology to 'dilution' and the loss of trainee apprenticeships led to "a loss of identity consciousness with the company and its products," said Mr Rooney. "Operators have to be encouraged to assume responsibility for product assurance, and workshop quality assurance is one of the important roles the foreman must fulfil."

All Rolls-Royce workers are now given quality control training, with each group attending a set course of one and a half hours a week spread over six weeks. They study cause and effect diagrams, pareto analysis, process capability studies and simple statistics.

"The quality control group is seen as a way of going through problems with a fine tooth comb, without adding to the burdens of management and engineers. When interdepartmental problems arise, the approach is to broaden communication by bringing the respective quality groups together in joint meetings," said Mr Rooney.

Group results are published on a formal basis with a wide circulation at shopfloor and management level and group members are given the opportunity to present projects to senior management.

"At first the groups concern themselves with problems of improved control of the local process and reduction of operator controllable defects," Jim Rooney explained.

Firmer grip

As a firmer grip is secured on these control problems, more elaborate projects are chosen by the group. As the group gains more experience it may find itself dealing with quality problems, caused by other sections of the factory, for instance, design engineering. In this case, recommendations are made by the group which are then taken up by the production and quality engineering departments.

Jim Rooney also stressed the importance of the foreman in operating a successful quality circle programme. "He has to select the problem to be discussed, devise ways and means to analyse and solve the problems, promote improvements and foster the concept and methods of statistical analysis."

"It is up to management to provide adequate training facilities, with budget support and the employment of professional tutors in quality control techniques. The groups should not be regarded as the prerogative of manufacture but should be extended to other departments including design, planning, accounting and administration."

Management he added also has to emphasise the importance of the groups. "Control of quality and effective attack on nonconformance costs will not be successful unless the groups are regarded as an integral part of the way we run our business."

Following the conference, Executant are offering a quality control consultancy to industry and aims to provide training for control group 'leaders'. The company has also formed QC Circle Society and is to organise of seminars on the technique. ■

New Technologies Put QC On the Production Line

By Robert R. Irving

A good place to look for the new inspection technologies is within aerospace itself where they are put in the hands of quality engineers.

Quality control can go just so far with such time-honored tools of its trade as micrometers, height gages and calipers. This fast-growing discipline now has at its disposal a veritable arsenal of new methodology designed to place testing and inspection on the same level as manufacturing itself.

It's about time? Yes, and none too soon, either.

The buzz words used to describe this new era are "quality technology." Evidence of the emergence of quality technology in the aerospace industry was illustrated last week at the American Society for Quality Control's annual technical conference in Atlanta. There were even signs that the automotive industry had jumped on the bandwagon as well.

Responsible for this new thrust in quality engineering is the technological explosion under way in the field of electronics. A good example is the development of computer-aided inspection (CAI), a new milestone in quality assurance.

The CAI approach is best epitomized by the computer-controlled coordinate measuring machine or CMM, a technique now being used by a number of major aerospace companies to measure parts which have been machined on numerically controlled machine tools.

According to J. Y. McClure, vice president, quality assurance, Fort Worth Division, General Dynamics Corp., the software programs controlling the movements and measurements of CMMs are being prepared by quality engineers.

Use of CMMs, he points out, has been estimated to result in a 4-to-1 improvement in productivity. These machines make possible a rapid and accurate comprehensive inspection of nearly every dimension of complex machined parts by the system's electronic measuring probe.

Harping on a similar theme, Roger G. Langevin, vice president of Intertek Services Corp., New Canaan, Conn., points to the interactive computer graphic system at Lockheed-Georgia Co. where most engineering designs are being released on microfilm-plot from the computer data base, thus eliminating the need for the vellum or Mylar master.

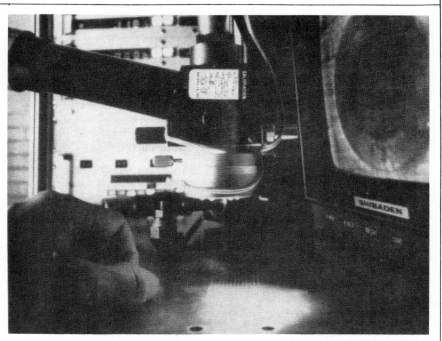
Closed-circuit television systems like this one at Bell Aerospace-Textron are designed to make inspection functions more reliable.

"A computer-controlled coordinate measuring system," he says, "is being used there to automatically measure parts and print out inspection results. The machine is programmed from either an actual part, an engineering print or from computer data-base information."

Mr. Langevin has recently completed a survey of the nation's top defense contractors in order to obtain information about the latest applications of quality technologies.

Closed-circuit television, he notes, is also being put to wider use by many contractors. For example, Bell Aerospace-Textron has added closed-circuit TV to an existing 3-axis coordinate measuring machine. This equipment inspects microscopic characteristics on intricate parts with close tolerances at magnifications of 80X.

"The accuracy is on the order of ±0.0003 in. for the entire system," observes Mr. Langevin, "with a repeatability of ±0.0002 in. Savings for the system, over conventional microscopic techniques, are on the order of 8 to 1."

The Defense Systems Division of Honeywell, he reports, has applied automatic non-contact gaging to several areas, including printed circuit board inspection, parts orientation and in inspection during assembly, gaging of features on piece parts and assemblies, and the detection of flaws on machined parts.

Honeywell has also developed a general-purpose high-speed microcomputer-based TV camera, visual inspection system and are constantly refining the inspection system's speed, resolution and data processing.

TV monitoring has also moved into the area of radiographic inspection. Mr. McClure describes one particular system of TV monitoring and video tape recording which includes a material handling system, an X-ray unit, film storage and transport mechanism in a radiation-shielded enclosure. It's all operated from a remote console panel.

"It has been calculated," says Mr. McClure, "that the combination of improved inspection methods, inspection manhour reduction and reduced material costs will provide a return-on-investment in approximately 14 months."

Photogrammetry is another technology moving into aerospace plants. In photogrammetry, two cameras are used to photograph an object from two vantage points. The photographs are evaluated on an optical reading device. The results are then fed into a computer that performs the calculations needed to yield precise dimensional measurement data.

Photogrammetry, concludes Mr. McClure, provides at least a 2.5-to-1 improvement in productivity over conventional dimensional measurement methods. "It also offers opportunities," he adds, "to expand the technology in such areas as detail part inspection, structural measurements of large components, and full-scale inspection of completed aircraft."

By replacing the photographic plates with a light source and sensors, photogrammetry could be developed into a real-time three-dimensional measurement system. It could then be used as a non-contact method of on-machine inspection.

From what was described at the ASQC meeting in Atlanta, it would appear that R&D must be really booming in the field of ultrasonics.

Mr. McClure points out that computerized ultrasonics is being applied to the detection of surface and subsurface flaws. With this approach ultrasonic energy is transmitted through the part being inspected, using water as a couplant.

"The energy is received, amplified and used," he says, "to drive a C-scan recorder for a permanent record of detected defects. This overall system has been shown to increase productivity by a factor of 4 to 1."

Future improvements are in store for this computer-driven ultrasonic inspection system. Eventually the pulse-echo technique will be included in the system. This will provide ultrasonic signal data along with scanner location data into a data acquisition computer. From the computer, the data can be displayed on a graphics terminal or it can be stored on magnetic tape.

At Lockheed-Georgia Co. a computer-controlled coordinate measuring machine is used to automatically measure parts and print-out results.

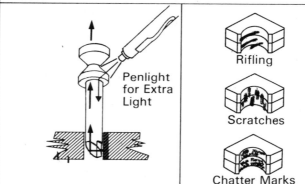

Sight Pipe Tool Examines Hole Surface Texture

"This advancement," predicts Mr. McClure, "will eliminate the bulky 1-1 C-scan chart and will enhance productivity even more."

According to Mr. Langevin, McDonnell Douglas has developed an ultrasonic system capable of scanning composites and metallic structures automatically and under full program control. The software has been designed so that part programs are "operator established." The operator simply moves the test probes over each corner or test boundary point on the first part and inputs to the computer accordingly.

McDonnell Douglas is now improving the graphics system so that the data can be reviewed prior to plotting. Also, a complete computer analysis of ultrasonic test response is in the offing. Thus, the computer will then be able to recommend, routinely, acceptance/rejection decisions for production parts.

Mr. Langevin also reports that Boeing has developed an ultrasonic sweep scanner for in-service detection of damage to large honeycomb structures such as radomes. In this setup, the transducers are mounted alternately in a "carpet sweeper" test scanner with visual defect indicators. Both the transducers and indicators are powered by a portable power supply.

This portable inspection device, with its 12 in. sweep width, has reduced test time at Boeing by 20 to 1.

Aerospace is also putting acoustic emission to good use. One such system has been developed by Grumman Aerospace Corp. for detecting and locating crack initiation and growth during severe fatigue cycling of elements of materials and aircraft production components.

Grumman is also said to have used acoustic emission monitoring to estimate the ultimate fatigue strength of composites.

At the Kennedy Space Center, says Mr. Langevin, specially designed acoustic detection sensors are being used for proof-load testing of high-temperature insulation tiles on the Space Shuttle. This is being done during the systems test phase.

The proof test assembly, built by the Shuttle Orbiter Division of Rockwell International, consists of acoustic sensors, a pneumatic controller, vacuum gages and an alignment device. The acoustic system has a detector transducer, pulse generator, recorder and calibration tile.

The tile to be tested is loaded in tension and the acoustic transducers in contact with the tile "listen" for adverse noise levels and transmit the information to the recorder. After the test, a compressive load is placed on the tile and its strain isolation pad to restore it to pre-test specifications.

The problem of sub-surface cracks, emanating from fastener holes in aircraft structures, has received a lot of attention in recent years. According to Mr. Langevin, Lockheed-Georgia Co. has developed a microcomputer system which uses output signal data from an eddy current probe in computer memory.

As the data points are read, the sequential pattern of the points is analyzed and compared to patterns known to be representative of crack "signatures." Whenever input data correspond to a crack signature, an appropriate visual or audible signal is emitted by the computer.

In the aerospace industry, notes Mr. McClure, the advancements in fluorescent penetrant inspection are found mainly in more sophisticated methods of moving the parts through the various pre-treatments. Here conveyors are designed to allow pre-penetrant etched parts to be processed without manual unloading and reloading for further chemical processing.

The parts are processed through an electrostatic penetrant spray booth, rinse/dry booths and are finally lowered into a black light inspection booth.

"This system," says Mr. McClure, "provides for penetrant inspection to be accomplished within normal manufacturing flow, thus maintaining production rates while, at the same time, reducing inspection manhours. In fact, savings ratios from 3 or 4 to 1 have been reported within the aerospace industry."

Of course, the use of the computer for quality data acquisition and preparation of management quality reports is commonplace in the aerospace industry. However, even though storage, retrieval and management reports have all been computerized, the floor-level entry is still a tedious, manual process.

Mr. McClure gives an example of how direct data entry is being done in connection with a computer-controlled coordinate measuring machine. In this instance a comprehensive, automated lot sampling system is provided for dimensional measurements of complex NC machined parts.

The new quality technologies are giving management the ammunition to speed up QC.

"With an automated CMM system," he explains, "sample size and acceptance/rejection tables of MIL-STD-414 are programmed into the computer. The CMM operator identifies to the computer the part number to be inspected and the number of parts in the lot. From thereon, the decisions required by MIL-STD-414 are handled by the computer system.

"The CMMs measure the required sample size and compare the results to the known variance data base and an accept/reject decision is made for that lot of parts. If a reject decision is made, the computer will print out those features which constituted the reject decision for screening on all of the remaining parts for that lot of parts."

Mr. McClure also sees automation coming to various calibration functions. In such a system, manual calibration functions will give way to microprocessors and mini-computers.

He also predicts that further technology acceleration of in-process control and monitoring operations will be directed toward complex structural panel fabrication made of graphite-epoxy composite materials. Already developed in this area is a technique for automatically analyzing composite materials using chemical characterization data.

One aerospace firm reports a 20-to-1 improvement in productivity using this approach.

"Complete automatic control of this type process," Mr. McClure remarks, "may be several years away, but progress toward this end is increasing rapidly.

"Since the computer will be used to fully control and monitor this process, the knowledge gained opens the door for computer application to many other similar processes for self-control such as fabrication of fiberglass components, application of coatings and finishes to parts, sealant and adhesive control, and various metallurgical processes."

As the aerospace industry works diligently to unify quality and manufacturing technologies, the automotive industry is attempting to do the same in certain areas. A very good example in automotive is in the torquing of bolts.

At the ASQC meeting, David R. Schwinn, staff quality engineer at the Ford Motor Co., discussed his company's experience with torque-rate control, a system whereby the fastener is both tightened and inspected simultaneously.

Ford Motor Co.'s first torque-rate control system was installed about two years ago in its Lorain Assembly Plant. Shortly thereafter, a second system was installed at Oakville.

Prior to purchase, this system from SPS Technologies went through a two-year investigation. Included in the investigation were system capability and reliability, provision for tool backup/failure, repairing and servicing.

The new system was chosen by Light Truck Engineering as the most cost-effective means of upgrading the steering gear-to-frame attachment (three 7/16-14 grade 9 bolts).

Since the installation of torque-rate control systems at Lorain and Oakville, points out Mr. Schwinn, more than 300,000 steering gears have been installed with no significant manufacturing problems reported.

Mr. Schwinn notes the following benefits:

1) Full utilization of bolt strength capacity.

2) Essentially 100 pct inspection of bolts for gross defects.

3) 100 pct inspection of the assembly process for cross threading, low or high torque, or wrong bolt.

4) Identification of repair required.

5) Reduced frequency of manual torque audits.

Item No. 5 represents the principal cost savings benefit of the SPS tool. Projects to date have returned about 100 pct on the investment.

In addition to SPS, the following companies also produce systems fea-

How the Computer Fits Into Automated Sampling

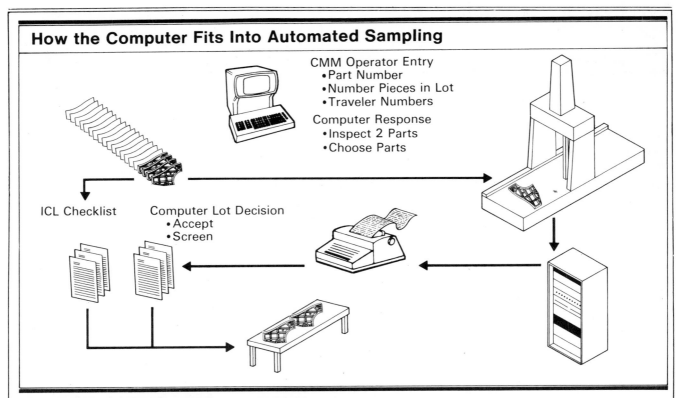

The quality visibility center at Northrup, Defense Systems Division, is used continually for corrective action and product assurance goals.

turing the torque-rate control concept: Ingersoll-Rand, Atlas-Copco, Stewart Warner and Rockwell International.

Other production installations outside Ford are:

1) Chevette rear axle bearing cap bolts are assembled using a 4-spindle unit operating at the General Motors plant in Detroit.

2) Deere & Co. has a 2-spindle unit installing connecting rod bolts on a large 6-cylinder tractor diesel engine. This customer is said to be satisfied with the improvement in joint quality realized from the method. They have eliminated the requirement of dropping the oil pan and retorquing the joints after engine hot test.

3) British Leyland has a 12-spindle unit installing cylinder head bolts on their Princess car line. In one of their advertisements, they state it is no longer necessary for an owner to bring his car in for retorquing of the cylinder head bolts.

It seems that many of these new technologies are giving rise to some re-thinking within quality assurance. In this connection, Mr. McClure of General Dynamics mentions a new concept—group technology—which is attracting considerable attention throughout industry.

"Currently," he explains, "manufacturing shops are usually laid out according to types of processes or manufacturing operations. Such a layout results in inefficient work flow, difficulty in scheduling work, restricted and complicated material handling, and difficult status reporting of work-in-process.

"Group technology is a means of reducing non-productive time. Under the group technology concept, shop layout is based on 'co-locating' all the manufacturing/quality tasks associated with a given family of parts. In this way, a work cell is formed to complete all the related manufacturing and inspection steps necessary to complete the part."

Some companies, Mr. McClure notes, have installed experimental work cells and all report outstanding cost benefits. "For example, some companies report 40 to 80 pct reductions in flow time and a few report 40 to 75 pct reductions in set-up time."

From the looks of things, it would appear that quality technology can play a major role in putting American productivity back on track once more. □

Computer-assisted quality control

The Ford Motor Co is applying a network of full-fledged computers to control quality during production of the company's new automatic overdrive transmission. The computer system directs objective product testing on the plant floor and recommends repair procedures for units deviating from design specifications.

by John R Coleman
Assistant Editor

The automatic overdrive (AOD) transmission, **Figure 1**, produced at Ford's Livonia, MI, transmission plant will play a major role in achieving company fleet mileage goals during the 1980s. The transmission assembly contains over 700 component parts. This design complexity makes it imperative that all parts meet strict manufacturing tolerances. Deviations from these limits could inhibit correct functioning of the transmission.

In order to insure product integrity throughout the AOD manufacturing process, the plant's quality control department has integrated conventional techniques (i.e., contact gaging frequencies, lot sampling etc) with objective computer-controlled testing procedures. Effectiveness of the entire QC system is monitored daily by a final quality audit (FQA) that samples the quality level of assembled AOD transmissions already accepted as good units.

Machining the AOD housing

The largest and most important component of the AOD transmission is the housing. It is an SAE 308 aluminum die casting, weighing approximately 30 lb and is completely machined on a five-unit transfer line. The transfer line sequence is a series of over 800 milling, drilling, tapping and boring operations

1. Cutaway view of the AOD shows internal complexity of the assembly. Tests indicate the AOD provides up to four additional miles per gallon in highway driving compared to non-AOD power trains. Fuel-efficiency improvement is achieved by using a gear ratio of 0.67:1 and by eliminating hydraulic slippage losses in the transmission's torque converter. The AOD has a split-torque path (60 percent mechanical and 40 percent hydraulic) in third gear and a full mechanical lock-up in overdrive.

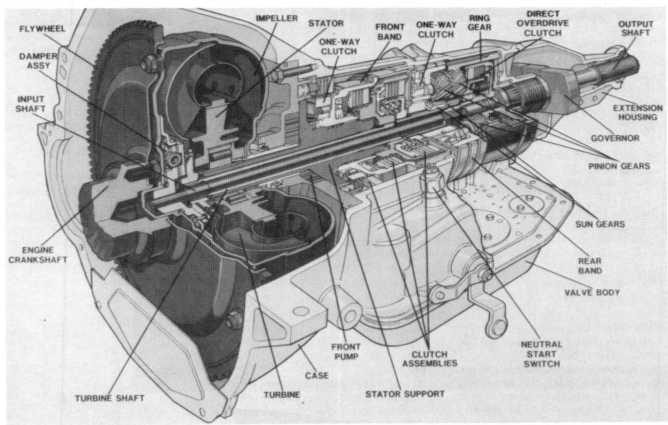

that remove a total of 1 to 1½ lb of aluminum.

Each of the transfer line units is controlled by a dedicated programmable controller. The PC's communicate among each other, thus allowing any combination of the machining sequence to be performed. This permits bypassing an operation that is shut down; however, housings with an omitted operation must be accumulated off-line and fed back into the system at the convenience of production management.

A team of inspectors and production job setters is responsible for maintaining gage frequency checks necessary to control the machining process, **Figure 2**. After all machining is completed, each housing is transferred through a liqui-deburring line where jet blasts of water at 5000 psi remove feather edges and burrs. The housings are then routed to a computerized air leak test line, **Figure 3**.

The leak test consists of pressurizing a housing at 15 psi while a computer monitors rate of pressure loss in valve areas, servo bores etc. Leaks are usually attributed to excessive die casting porosity exposed by machining.

The computer divides a housing into five sections, each receiving a preassigned internal leak rate specification. The limits were developed according to how critical a section's relationship is to the function of the transmission.

Rejected housings are routed to a repair area where an anaerobic coating is sprayed on sections not complying with the leak specification. The coating is allowed to cure for 6 hr before the housing is resubmitted to the air test line.

At random, housings rejected by the air test computer are removed from the repair loop and taken to a water test area. Here, a housing is submerged, pressurized and observed for leaks. This is an audit of the computer's decision performance.

Housings that pass the air test are transferred by overhead conveyor to final assembly. Other subcomponents (i.e., valve body, sun gears, direct drive shaft etc) are machined in other parts of the plant and brought together with the housing at final assembly.

2. Machined AOD housing is clamped on a fixture plate for dimensional checks. The plate contains embedded probes that read surface flatness. Flatness is determined by the height of each column (red) shown on monitors (background).

3. Computerized leak test line from Wilson Machine, Saginaw, MI, is a palletized nonsynchronous system designed to automatically divert pallets into one of five test stations. Each pallet contains a flag that signals a gating mechanism whether a transmission is accepted or rejected. Maximum line rate is 300 units/hr.

The assembly line is a 600-ft-circumference carousel containing three inspection stations. Each station reviews all preceding assembly operations. Transmissions requiring rework are directed off the assembly line to a repair loop. Units in a repair loop will stay there until all defects are corrected. Fully assembled units are routed to a computerized test stand area.

Computer-directed testing

The test stand area employs one of the largest plant floor installations of full-

Computer-assisted quality control

...continued

4. One of 37 computerized test stands that objectively checks each AOD transmission. The dedicated computer cycles the stand through various phases designed to simulate installation in a vehicle.

5. Transmissions rejected by the test stand check are diverted to a repair area. Here, a repairman requests information on a particular rejected transmission from the computer system. The system responds with why the unit was rejected and suggested areas to look at in order to fix the problem.

fledged computers in the country, **Figure 4**. There are 37 test stands, each with its own dedicated computer. These computers form a network supervised by four main control computers. All the computers are of the Texas Instruments 990-10 Series and are programmed using assembler language.

In the past, a test stand operator would monitor meters and gages to determine a transmission's quality level. This was considered subjective testing since it required a human to decide if a unit was performing adequately. Now, a computer objectively determines the product's quality without human intervention. If a transmission is rejected, the operator cannot override the decision; the unit is automatically routed to a repair area. However, if a computer determines that a transmission is acceptable and the operator observes a fault, he can override the computer's decision.

Currently, the test procedure starts by manually loading a transmission into a test stand fixture. At this point the computer begins a vehicle simulation. The test sequence involves a complicated mix of up-shift patterns, down-shift patterns and parking mechanism checks. Also, the computer simulates a condition where the vehicle's accelerator and brake are fully depressed while the transmission is engaged in forward and then reverse. This is an extremely adverse test, since the transmission must absorb the developed energy.

Test sequencing takes approximately 3 min. Ten analog signals, generated by strategically placed transducers, are monitored by the computer 20 times/sec. These signals represent input/output speeds, input/output torques, hydraulic pressures in various clutch packs etc.

The computer uses this information to determine if the transmission is functionally acceptable. Information from each test stand computer is transferred back to the main control computers for compilation and analysis. This analysis can be used for adverse trend detection. A report is generated by the computer system detailing reject modes. This data is used to immediately correct problems on the plant floor before there is a significant build-up of rejected material.

Rejected units are automatically routed to a repair area, **Figure 5**. The AOD's complex design makes it difficult to pinpoint areas that are causing rejections. The computer system is used to expedite this process.

A repairman may request information on a rejected unit by inputting its serial and fixture code. The main control computers respond via CRT with the following data: (1) actual test readings compared to test specifications and (2) five suggestions on how to fix the problem. Suggestions are ranked by their probability for success. Repair data is updated constantly by success rates experienced on the floor. If a suggested repair is not effective, it will lose ranking among the top five possibilities.

When repairs are completed, the main control computers track the transmission back to the test stand area where it repeats the functional test sequence. This computer assist allows a repairman to zero in on a problem much more rapidly than was previously possible, thus utilizing his time to a greater degree.

Assemblies accepted by the test stand computers must pass an external leak test, **Figure 6**, before they are sent to the shipping dock.

Final quality audit

To check total QC integrity, a sample of 20 transmissions per day (against a build of 1600) are pulled from the shipping dock. Fourteen of the units are completely disassembled and each part is inspected against a partprint. In addition, six units are installed in vehicles and road tested on a test track located on the plant's grounds, **Figure 7**.

The FQA detects if the QC system has short circuited, whether gages or test procedures are misleading or if there are problems not related to gaging or testing procedures. Any detected substandard conditions will trigger a 100 percent re-verification of the integrity of all units on the plant floor. Of course, a performance failure during FQA could tie up an entire production build. This is the driving force behind Ford's management philosophy of "do it right the first time." ■

6. Fully assembled transmissions that have been accepted by the test stands are routed to a water test that checks for any external leaks. Units that pass this checkpoint are ready to be shipped to an assembly plant for installation into a vehicle.

7. Aerial view of Ford's Livonia transmission plant shows the test track (arrow) used for part of the final quality audit (FQA) procedure. The track has a 0.8 mile circumference. Construction of the plant was completed in 1952, and production of automatic transmissions began in 1953. In April 1978, a 318,000-sq-ft-expansion was completed for production of the AOD transmission, bringing the total plant manufacturing area to 2,945,385 sq ft. Plant-wide manufacturing operations are performed by over 1950 machines and consume over 2,300,000 lb of raw material per day. Production materials are delivered each week by 43 railroad cars and 535 cargo-carrying trucks. Like a small city, the plant consumes approximately 14,000,000 kWhr of electricity per month, enough to meet the needs of 20,000 homes; 2350 tons of coal; 30,400,000 gal of water; 68,200,000 cu ft of natural gas; and 768,000,000 cu ft of compressed air. There are approximately 5000 employees at the Livonia facility.

Operator's Involvement in Company QC Programme

by Ravi L. Kirloskar
Chairman and Managing Director
Kirloskar Electric Co. Ltd.
Bangalore

In this article the benefits and advantages secured by involving the operators in the company's QC programme are described by the chairman of a company producing electric motors, indicating also the steps taken in this behalf.

There is an increasing realisation that in the field of export, qualty and reliability of products are the most important considerations in achieving sustained sales. However, these considerations can no more be confined to exports only as they have become equally vital for domestic marketing. In our country too, there has been a greater awareness and understanding of quality as evidenced by the customers' growing demands for higher quality requirements. In this context management recognises that to retain the dominant position in the market amidst sustained competition, the best means is through manufacturing quality products at economic cost. It is realised also that this cannot be achieved except through a company-wide QC programme which involves the adoption of systematic quality control procedures that aid in building quality into the product. This, in turn, demands active support by top management as well as involvement at all levels, including operators, in the company QC programme. To reap the full potential for improvements through the QC programme, it is also found that such involvement of operators in addition to other levels is vital, though it is no easy task to evoke such response from them. This article brings out our experiences in tackling this problem in the Kirloskar Electric Company.

PREVENTIVE OUTLOOK

While no doubt right from the inception of the factory in 1948, we are having an Inspection and Testing Cell, the increasingly higher quality requirements of customers have necessitated updating the past quality concepts, practices and techniques to meet these demands. This has resulted in adopting the modern concept of quality control with its primary stress on prevention of defects rather than relying only on an elaborate sorting inspection of outgoing products for assuring quality. By its very nature, prevention-oriented efforts call for the involvement of all the sub-systems affecting quality, such as purchase, design, manufacturing, marketing, etc. In view of the continuous growth of the organisation over the years, getting across these ideas to the various levels, including shop floor workers, has been a major educational exercise in communications due to the large numbers involved and the well-known human resistance to change.

TRAINING

To overcome these inherent difficulties and to act as a catalyst in changing the attitude of the personnel, it was considered desirable to call upon the aid of an external agency. Accordingly, this task was entrusted to the SQC Unit, Bangalore, of Indian Statistical Institute, in 1973. Agreeing with the view of the management that quality is basically an attitude of mind which has to be cultivated on the right lines at all levels, the first year of their association with us was devoted to exposure of senior management and middle management as well as supervisors, to the latest concepts of prevention-oriented company-wide quality control programme through a series of inplant training courses. During the second year, attention was directed to secure all-around acceptance of the new concepts. To supplement these efforts in the task of training, management spared no efforts in sponsoring personnel to various inplant and outside courses on QC conducted by different agencies, as can be seen from Annexure 1.

APPLICATIONS AND DISCUSSIONS

A special feature of the in-plant intensive programme for middle management organised by the SQC Unit was the combination of theory with practice by way of project work by the participants. As a sequel, in order to promote company-wide applications, QC groups were constituted from among the production personnel in each project line for machine shop, sub-assembly, winding and assembly operations, quality rating, etc. The highlights of their work were presented at periodical intervals in in-plant seminars in the presence of the top management. This had a great impact, so much so that subsequent years have been devoted to in-depth special investigations under the guidance of SQC Unit by the various participants in different sections.

OPERATOR INVOLVEMENT

While a number of gains were reported through these studies, a misapprehension appeared to develop in the minds of workers when feed-back from several studies involved operator-wise performance in addition to machine-wise. To allay such misapprehensions, the advantages of modern QC were explained by the management to the workers' representatives during patient and detailed discussion with them. While as a result the workers agreed to give a fair trial to the QC system, this, it was felt, was an opportune time to actively involve them also in the company QC programme. As a first step, it was decided as a matter of strategy to take one product line for concentrated effort.

To create the necessary receptive climate for induction of operators into the company QC programme, specially designed training programmes in the regional language (Kannada) were organised for them. An explanatory booklet in Kannada was also prepared. The operators were drawn in batches, each consisting of about 20 persons. For each batch the training programme was conducted for 12-15 hours spread over a period of two weeks.

OPERATORS RESOLVE QUALITY PROBLEMS

In these courses, apart from exposure to basic QC concepts, and simple but powerful QC methods the operators were apprised of the shop quality problems. Further, different types of failures at the factory as well as at customers' end were brought home to them. They were then invited to come out with their own assessment of reasons for various quality problems. In order to ensure that they spelt out these reasons without any inhibition or reservations, it was so arranged that they had several meetings on the shop floor with SQC Unit Specialists without the presence of their supervisory personnel.

During these meetings the operators made hundreds of observations bearing on product quality arising from defects of various origins such as design, materials, sub-contracts, tool room, processes, etc. They also offered many effective solutions like use of additional/improved fixtures, changes in drawing to debug process defects, use of better materials and improved processes. Based on these, a suitable plan of action relating to their own shop as well as design, tool room, feeder shop, etc., was drawn up. Progressive implementation yielded sufficient heartening results. Encouraged by the good response, a firm decision was taken to extend the scheme progressively to cover the entire plant. So far, two more divisions have been covered and 180 operators from all the three divisions have been trained.

In this context, special mention has to be made of an application which goes much to the credit of workers. Test bed rejections in one division were chronically high, defying all efforts of management. When workers were apprised of the problem, though with a caution that this may be beyond them, they took a spirited challenge that they could show results in this section too. The following table serves to give a picture of the impressive results achieved in this as well as other sections in 1976-77 (when for the first time workers were inducted into the QC programme) as against the figures for earlier years:

PERCENT REJECTIONS IN A DIVISION

Sl. No.	Particulars	1973-74	1974-75	1975-76	1976-77
1.	Test Bed	11.3	19.6	16.8	5.1
2.	Control Box Assembly	11.2	3.3	4.8	1.7
3.	DC Frame	18.8	7.7	5.7	3.2
4.	Rotor Winding	12.6	1.5	0.8	0.6
5.	Stator Winding	5.2	4.0	3.4	0.6
6.	Final Assembly	7.4	3.3	8.0	2.6

As regards the other two divisions so far covered under this programme, there has been, if anything, even better response from the operators who have likewise come forward with a large number of causes and solutions for the various quality problems. Efforts are under progress in evolving suitable plans for action in each case, taking into account all relevant factors.

OVERALL GAINS

To sum up, the highlights of operators' involvement in the factory QC programme are as follows:

(a) Quality improvement through better process control is becoming part and parcel of the routine QC activities.

(b) Breakthrough has become possible in improving the performance of certain machines, which were once considered impossible to improve.

(c) Accuracy of components is being assured by paying attention to the accuracy of set-ups.

(d) Operators are deriving greater job satisfaction.

(e) There is a pronounced decrease in customer complaints.

(f) Operators and supervisors who were once suspicious and resistant regarding QC concepts, are becoming the votaries of the same.

(g) In our efforts to hold the price line for our products, the company has been helped in partly absorbing the rising wages and material costs by the savings achieved in quality losses and increased productivity.

In conclusion, while results achieved so far are quite promising, it has to be stressed that the consolidation of these gains is more onerous than registering new successes. This important consideration has to be taken into account in planning for the desirable objective of further extension of the scheme to cover other divisions.

ANNEXURE 1

Coverage of Personnel in Training Programmes

Sl. No.	Courses	Senior Management	Middle Management	Group Leader & Supervisor	Operators
1.	In-plant programmes by SQC Unit, Bangalore	20	60	75	180
2.	In-plant programmes by National Productivity Council	...	30	30	...
3.	External programmes:				
a.	Six-month evening course by SQC Unit, Bangalore	...	31	...	...
b.	8-week evening course by IAQR	...	...	35	...
c.	1-year and short-term courses by Foremen Training Institute		62	...	...

ABOUT THE AUTHOR:

Ravi L. Kirloskar, Chairman and Managing Director, Kirloskar Electric Company, Limited, Bangalore, graduated from Worcester Polytechnic (USA) and worked with Westinghouse Electric Corporation. He is Founder Director of the joint venture, Indo-Malayasia Engineering Company, Berhad and Director of several other Indian Companies. Served as President, Institute of Management, Bangalore and Chairman, Indian Electrical Manufacturers Association. Won the Robert H. Goddard Award of Worcester Polytechnic in 1974 for outstanding professional achievement. Fellow IEEE and chairman, Bangalore Section; Fellow, Institution of Engineers (India); President, IAQR (National Body).

INDEX

A

Absenteeism, 116, 230
Accident investigation, 180
Accounting, 44, 230
Acoustic
    emissions, 232
    sensors, 233
Adequacy audit, 65
Adhesive control, 233
Administration, 47, 230
Aerospace, 231
Air
    leak test line, 236
    test computer, 236
Alignment, 233
Amplication, 189
AOD, See: Automatic overdrive transmission
Applications
    automatic warehousing, 92, 222
    awareness, 17-26
    customer service, 8, 65
    electronics, 191, 231
    financial, 71
    injection mold control, 222
    operators, 4, 239-243
    planning, 42, 66, 132, 160, 230
    psychology, 121-127
    technological, 10-13, 71, 212-213, 215, 231-234
    thermal analysis, 189
Appraisal costs, 47-49, 79
Aptitude tests, 125
Army Tactical Data Systems, 197
Aspects
    health, 214
    international, 10-13, 212-213, 214-218
    multinational, 207-211
    national, 214, 218
    safety, 45, 160, 168, 179, 212, 214
Astronomy, 121
Audit, 9, 27, 48, 61-64, 65-70, 71-73, 161, 238
Authors
    Amsden, Davida M., 225-229
    Amsden, Robert T., 225-229
    Bajaria, Hans J., 53-58, 131-134
    Bell, A.C., 141-147
    Birbara, Philip J., 27-33
    Blank, Lee, 79-84
    Brecker, J.N., 3-6
    Cameron, P.J., 159-164
    Coleman, John R., 235-238
    Cooper, James E., 74-76
    Davis, C.R., 17-26
    Dellon, Robert E., 61-64
    den Boer, John H., 47-49
    Dilworth, G.F., 151-158
    Edmonds, C.R., 135-140
    Freund, Richard A., 10-13, 212-213, 214-218
    Fromson, R.E., 3-6
    Gossard, D.C., 141-147
    Irving, Robert R., 231-234
    Jones, J.R.D., 159-164
    Kaushagan, W. Maurice, 114-120
    Keezer, Edward I., 197-203
    Kirloskar, Ravi L., 239-243
    Koenig, Myron, 189-190
    Lynch, Richard B., 167-178
    McWhinnie, John R., 34-40
    Miller, J.L., 221-224
    Mills, Chas. A., 7-9
    Murray, David J., 41-46
    Negro, Fernando D., 207-211
    Peach, Paul, 121-127
    Puri, Subhash C., 34-40
    Ryzeszotarski, Albin J., 65-70
    Sade, Norman G., 182-185
    Shanmugam, K.M., 104-113
    Solorzano, Jorge, 79-84
    Suh, N.P., 141-147
    Takanashi, Masahide, 191-196
    Thomas, E.F., 87-96
    Veen, B., 97-103
    Walton, C.W., 179-181
    Wilborn, Walter, 71-73
Automated sampling, 234
Automatic
    data requisition system, 140
    overdrive transmission, 235
    test, 223
    warehousing, 43, 222
Axiomatic approach, 141-147

B

Background, 167
Balance mechanism, 190
Baseline, 200
Benefits, 62, 71, 79, 239
Brainstorming, 226
Brake linings, 226
Budget, 72

C

CAI, See: Computer-aided inspection
Calibration, 233
Calipers, 231
Calorimetry, 190
Canada, 223
Casting waxes, 189-190
Cause-and-effect diagram, 56, 227
Certification, 213, 217
Checklists, 73
Chemical, 222
Chip cutting, 5
Closed
    circuit television, 231
    loop feedback system, 225
CMM, See: Coordinate measuring machine
Coatings, 233, 236
Communication, 111, 131
Compliance audit, 65

Computer, 135, 160, 212, 231, 236
    aided inspection, 231
    assisted, 235-238
    capability matrix, 136
    directed testing, 236
Conditional response, 126
Configuration management, 199
Conformance, 71
Consequential costs, 79
Considerations
    design, 3, 8, 159-164
    methodological, 37
    process/inspection, 4
Consumer reports, 39
Contingency tables, 55
Controller, 221-224
Control system, 139
Cooperation, 212-213
Coordinate measuring machine, 231
Corollaries, 145
Correlations analysis, 55
Costs, 39, 47-49, 79-84, 138, 170, 230, 233, 239
Criticism, 111
Customer service, 8, 65
Cycle switch test system, 139

D

Data
    back-up, 48
    collection, 81, 223
    confirmation, 48
    management, 82
    requisition system, 140
Defense contract, 200
Department of Defense, 198
Design, 3, 8, 12, 28, 41, 47, 55, 132, 159-164, 171, 230
Development, 28, 44, 45, 191, 196
Differential
    temperature monitors, 190
    thermal analysis, 189
    transformer, 190
Discretionary costs, 79
Distribution, 43
DTA, See: Differential thermal analysis

E

Electrodes, 190
Electronics, 191, 231
Emissions, 232
Engine, 116
Engineering, 44, 81, 114, 151-158, 170, 179
England, 114, 118
Environment, 45, 212, 214
Equipment, 81
Evaluation, 109
Evolution, 208
External failure costs, 80

F

Fabrication, 233
Factory, 104, 117
Failure, 71, 83, 179, 195

Federal standards, 184
Feedback, 119
Finishes, 233
Finland, 115
Fishbone chart, 227
Food processing, 222
Forecast, 214
Foreman, 4, 225
Function, 34, 79, 168
Future actions, 56, 84

G

Gages, 233
Gains, 242
Germany, 114, 117, 124
Goals, 19, 212-213
Goal setting, 92
Government, 72, 151, 197
Group
    discussion, 108
    leaders, 109
    suggestions, 108
    technology, 234

H

Handling system, See: Material handling system
Hardware, 137, 212
Hawthorn effect, 97
Health, 214
Height gages, 231
Heisenberg Principle, 121
Hong Kong, 212
Human
    factors, 215
    relations, 92
Hypothesis testing, 55
Hypothetical axioms, 143

I

Implementation, 11, 26, 61, 65, 71, 88, 138, 171, 210, 212, 225, 228, 241
Implications, 12, 215
Inauguration, 107
Incentive plans, 88
Index
    failure, 83
    prevention, 80, 82, 239
    productivity, 74-76, 114-120, 131-134, 135-140
    scrap, 83
    worklife, 134
India, 104
Indicators, 34-40
Industrial engineer, 81
Injection mold control, 222
In-plant, 240
Inspection, 4, 12, 29, 38, 49, 62, 74-76, 79, 231
Inspector, 5, 74
Integrated circuit, 191-196, 224
Integration, 97-103
Integrity assurance, 5
Interactions, 131-134

Internal
    failure costs, 79
    leak rate, 236
International
    aspects, 214-218
    cooperation, 10-13, 212-213
Investigation, 180
Investment casting waxes, 189-190
Involvement, 239-243

J

Japan, 97, 193, 212, 215, 230
Job
    enrichment, 90
    satisfaction, 87, 89, 97, 114
    setters, 236

L

Legal, 43
Legislation, 71
Liability, See: Product liability
Linear testers, 223
Liqui-deburring, 236
LVDT, See: Differential transformer

M

Magnetic tape, 232
Maintainability, 30
Maintenance, 230
Management, 17-26, 27-33, 34-40, 61-64, 79-84, 90, 103, 131, 179, 197, 199, 214, 233
Manitoba, 71
Manual calibration, 233
Manufacturing, 8, 18, 41, 47, 141-147, 170, 180, 221-224, 239
Marketing, 8, 42, 170, 180, 239
Material handling system, 232
Measurement, 4, 12, 213, 231
Microcomputer, 135, 221-224, 231
Microfilm, 231
Micrometers, 135, 231
Microprocessors, 135, 233
Minicomputers, 221-224, 233
Modeling, 202
Money, 88, 123
Motivation, 22, 87-96, 97-103, 115, 122, 215
Multi-
    divisional, 48
    national, 207-211
    plant, 47

N

National, 214-218
National Bureau of Standards, 5
Non-conforming material, 8
Norway, 114
Nuclear, 63, 65, 153, 161

O

Operator, 4, 239-243
Opposition, 111
Optical reading device, 232
Optimization, 141
Organization chart, 48
Organized labor, 225
Overdrive transmission, 235

P

Packaging, 44
Paper, 222
Pareto, 57, 225, 226, 230
Participative management, 17-26, 90, 197
Patrol inspection, 74-76
Peer group, 123
Performance, 91
Personnel, 19, 20, 22, 43, 45, 61, 79, 126, 152, 179, 240, 243
Peter Principle, 94
Philosophy, 169
Photogrammetry, 232
Photoreceptor, 139
Physics, 121
Planning, 3-6, 42, 66, 132, 160, 230
Plant, 44, 47, 116, 154
Pneumatic controller, 233
Portable inspection, 232
Post accident, 180
Prevention
    costs, 80
    index, 82
Preventive, 239
Probe, 189
Problems, 24, 35, 58, 74, 80, 99, 122, 209, 226, 241
Process control, 224
Product
    awareness, 17-26
    certification, 213, 217
    control, 173
    defense, 179-181, 182-185
    development, 28, 44, 45, 191, 196
    effectiveness, 207
    inspection, 4, 12, 29, 38, 49, 62, 74-76, 79, 231
    integrity, 5, 7-9
    liability, 49, 80, 167-178, 179-181, 182
    planning, 3-6, 42, 66, 132, 160, 230
    quality, 3-6, 17-26, 207-211
    reevaluation, 182
    safety, 45, 160, 168, 179, 212, 214
Production
    lines, 192, 231-234
    lots, 192
Productivity, 74-76, 87, 89, 105, 114-120, 131-134, 135-140, 231, 234
Program development, 47-49
Programmable controllers, 221-224, 236
Profit, 74-76
Promotion, 79, 106
Proposals, 99
Psychology, 121-127
Pulp industries, 222
Punishment, 89
Purchasing, 20, 43

Q

Qualitative methods, 56
Quality of life, 114-120
Quantitative methods, 53

R

Records, 47
Reevaluation, 182
Regression analysis, 55
Rejections, 241
Reliability, 30, 192
Reports, 39, 47, 72
Restatement of Torts, 179
Reversals, 39
Rolls-Royce, 230

S

Safeguards, 20
Safety, 45, 160, 168, 179, 212, 214
Sales, 20, 39, 42
Sampling, 234
Scotland, 164
Scrap
   function, 83
   index, 83
Sealant, 233
Self-inspection, 111
Semiconductor, 191, 193
Sensors, 233
Service, 8, 65, 192
Shortcomings, 87-96
Simulation modeling, 202
Singapore, 212
Small business, 71-73
Software, 137, 197-203, 212, 231, 232
South Korea, 212
Soviet Union, 215
Standard cost, 82
Standardization, 10, 217
Standards, 12, 213, 217
Statistical
   analysis, 218
   quality control, 53
Steering committee, 72
Strategy, 17-26, 71
Surveys, 71, 93
Sweden, 114, 118, 124
Systomation in-circuit testers, 223

T

TA, See: Thermal analysis
Tactical Data Systems, See: Army Tactical Data Systems
Tagging, 49
Taiwan, 212, 223
Tape, 232
Targets, 109
Taylor system, 230
Teaching, 225
Team, 27, 67, 111, 225

Technology, 10-13, 71, 212-213, 215, 231-234
Temperature monitors, 190
Tennessee Valley Authority, 151
Testers, 223
Tests, 29, 42, 48, 125, 139, 201, 212, 231, 235, 236
Thermal analysis, 189
Thermocouple junction, 190
Time allocations, 82
Torquing, 233
Training, 45, 61, 74, 81, 99, 108, 240
Transducer, 189, 232, 237
Transfer line, 236
Transformer, 190
Transmission, 235
Troubleshooting, 81, 137
Turnkey, 138
Turnover, 82
TVA, See: Tennessee Valley Authority

U

Ultrasonic inspection system, 232
Unions, 93
USSR, See: Soviet Union

V

Vacuum gages, 233
Vendor relations, 81
Videotape, 232
Violation, 69
Voluntary participation, 225

W

Wages, 88
Warehousing, 43, 222
Waxes, 189-190
Working conditions, 92
Work-in-process control system, 139
Worklife, 134

X

X-ray unit, 232

Y

Youth, 93, 123

Z

Zero defects, 104-113, 230